T0227984

MICROSCOPY HANDBOOKS 42

Introduction to Light Microscopy

Royal Microscopical Society MICROSCOPY HANDBOOKS

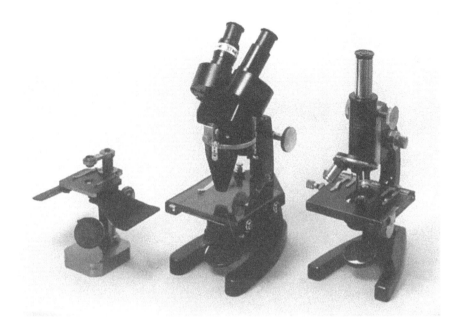

Frontispiece

Types of microscope. On the left is a simple microscope. Although this small Zeiss instrument dates from the 1890s, it is still perfectly adequate in use and many are still used today as dissecting microscopes. The heavy brass base carries a pillar with rack and pinion focusing for the lens-arm, which will rotate across the stage. The mirror is double-sided and the hand-rests facilitate sustained fine work. The small lens is divisible, to give powers varying from $5\times$ to $40\times$. In the middle is a Cooke, Troughton and Simms 'Greenough-type' binocular stereoscopic microscope, available for work with either or both transmitted or reflected illumination. The instrument is 345 mm high as pictured. On the right is a 1930s Zeiss laboratory-type compound microscope; it is this type which is usually meant when the word 'microscope' is used without qualification. Instruments like this are still in use in many laboratories today and are capable of much routine work.

Introduction to Light Microscopy

Savile Bradbury
Oxford, UK

Brian Bracegirdle
Cold Aston,
Cheltenham, UK

In association with the Royal Microscopical Society

© BIOS Scientific Publishers, 1998

First published 1998

Transferred to Digital Printing 2005

All rights reserved. No part of this book may be reproduced or transmitted, in any form or by any means, without permission.

A CIP catalogue record for this book is available from the British Library.

ISBN 1 859961 21 5

BIOS Scientific Publishers Limited,
9 Newtec Place, Magdalen Road, Oxford OX4 1RE, UK.
Tel: +44 (0)1865 726286, Fax: +44 (0)1865 246823
World Wide Web home page: http://www.bios.co.uk/

DISTRIBUTORS

Australia and New Zealand
  Blackwell Science Asia
  54 University Street
  Carlton, South Victoria 3053

India
  Viva Books Private Limited
  4325/3 Ansari Road, Daryaganj
  New Delhi 110002

Published in the United States of America, its dependent territories and Canada by Springer-Verlag New York Inc., 175 Fifth Avenue, New York, NY 10010-7858, in association with BIOS Scientific Publishers Ltd.

Published in Hong Kong, Taiwan, Singapore, Thailand, Cambodia, Korea, The Philippines, Indonesia, The People's Republic of China, Brunei, Laos, Malaysia, Macau and Vietnam by Springer-Verlag Singapore Pte Ltd, 1 Tannery Road, Singapore 347719, in association with BIOS Scientific Publishers Ltd.

Production Editor: John Leonard.
Typeset by Poole Typesetting (Wessex) Ltd, Bournemouth, UK.

Front cover: Photomicrograph of an unstained section of wood viewed by the use of Rheinberg illumination.

Contents

Preface

It is not the object-glass at the one end, but the head at the other, that constitutes the value of the microscope.

(Edward Forbes, 1815–1854)

Introduction to Light Microscopy is so-called because it is the basic work on the light microscope in the series of RMS Handbooks. It is a totally rewritten successor to Savile Bradbury's original No. 1 Handbook (*An Introduction to the Optical Microscope*) first published in 1984, which has been reprinted several times.

The present book offers fully illustrated advice on choosing and using the various kinds of microscopes in use today, and enough theory to allow users to come to understand why they are recommended to follow certain procedures, and avoid others! More specialized books in the series are referred to as appropriate, for a wealth of wisdom and experience is contained in them as a group.

It is, unfortunately, only too easy to get some sort of image with any sort of microscope, and only too easy thus to misinterpret the nature of what is being investigated. It is very little more difficult to get the best possible image from an instrument, and it soon becomes rapid second nature to set up the microscope correctly. It is our aim to stimulate microscopists of all kinds to do just this, and literally to enjoy the satisfaction which results.

We are most grateful to Drs P.J. Evennett and C. Hammond for reading the manuscript and providing very valuable and constructive criticism – their advice has been a great help to us and it has undoubtedly improved this book very considerably.

Savile Bradbury
Brian Bracegirdle

Safety

Attention to safety aspects is an integral part of all laboratory procedures, and both the Health and Safety at Work Act and the COSHH regulations impose legal requirements on those persons planning or carrying out such procedures.

In this and other Handbooks every effort has been made to ensure that the recipes, formulae and practical procedures are accurate and safe. However, it remains the responsibility of the reader to ensure that the procedures which are followed are carried out in a safe manner and that all necessary COSHH requirements have been looked up and implemented. Any specific safety instructions relating to items of laboratory equipment must also be followed.

1 Introduction

1.1 Why use a microscope?

Objects of interest to us exist in a very wide range of sizes (see *Figure 1.1*) but the unaided human eye is unable to see the smaller ones because its performance is limited. We rely very much on vision for the largest part of our information about the world about us, and the ability to see small things has come to play a large part in science and technology with consequences for daily life. Many documents are still stored in the form of microfiches which need magnification to allow them to be read. Microscopes are of value in fighting disease, for it happens that many parasites are small and many cellular components are smaller still, and to see both is to begin to *understand* both better. This simple statement conceals a vast amount of current scientific enterprise throughout the world, generally in specialist laboratories, and the accumulation of details of the most intimate workings of the bodies of all living things. Again, much surgery is now carried out with such precision that microscopes are needed in its execution.

Another area of microscope usage is in *design* and *manufacturing*, allowing smaller and smaller components, often in microelectronics, to be fabricated and assembled. Printed circuits in calculators, radios, televisions and computers are becoming ever smaller; these require low-power microscopes to help in their design. The use of microscopes in industry means that people who would never ordinarily have come into contact with a microscope now use one as a matter of routine every day. A hundred years ago the microscope had an important place in recreational activities: people studied minute structures and/or organisms for pleasure, appreciating the true beauty of form; this is still true today, even with the competition of television. Amateur microscopists still often make useful contributions to science especially in natural history. This introduction to the light microscope has been written for users of all categories. Many users have little understanding of how their microscope works. This has two consequences: one is that an inadequate

1

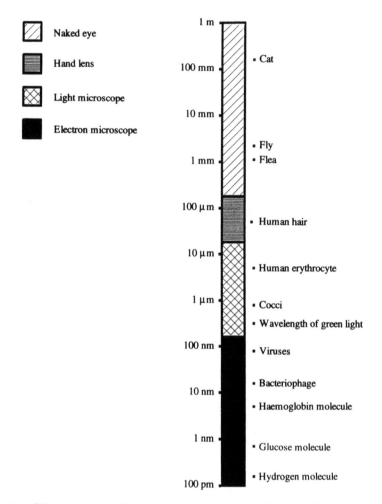

Figure 1.1. Diagram to show the relative sizes of a range of objects. The figures on the left of the bar chart indicate the size on a logarithmic scale, whilst on the right the approximate size of the object is indicated.

image for a particular purpose is sometimes obtained, and so the structure of the object may not be revealed. Lack of understanding of microscope usage, especially where high magnifications are used, also means that it is only too easy to produce an image which is a misleading representation of the object, so that totally wrong conclusions as to its nature may be reached. We hope that this book will help provide enough understanding of microscope theory to allow users to set up their instruments correctly and obtain the best possible images from them.

The limitations of the eye in its ability to see fine detail ('acuity' of vision) are due partly to the optical properties of the cornea and lens but also to the cellular nature of the light-sensitive retina. For two points in an image to be seen as separate it is obvious that their images must fall on at least two cells of the retina. The size of the image projected on to the cells ('cones') of the most sensitive area of the retina (the macula) is thus one factor which affects the amount of detail which may be perceived. Moving an object closer to the eye has the effect of increasing the size of the image on the retina – making the object 'look' bigger. But, because the unaided human eye cannot focus on objects closer than about 250 mm (often called the 'nearest distance of distinct vision'), there is a limit to the extent to which detail perception can be increased by this means. To see more detail, the object must in effect be brought 'closer' to the eye than allowed by the nearest distance of distinct vision; the instrument used to achieve this is the microscope. The microscope has revealed a whole new world since its invention around the beginning of the seventeenth century and new forms of microscope (such as the confocal instrument) are still being introduced. In this book we shall be concerned only with the basic light microscope.

1.2 What is a microscope?

The term 'microscope', strictly speaking, includes not only instruments fitted with separate objective and eyepiece lenses ('compound microscopes'), but also those composed of a single lens, even though this may have several separate components within a single mount. These single lens microscopes may be used either with or without devices for holding them and the objects to be examined. If the lens is fixed and there is a stage to carry the object under study, the instrument is usually called a 'simple' or 'dissecting microscope'; if there is no mount then we speak of a 'hand lens' or 'magnifier'. The hand lens will be considered in Chapter 2; nowadays such lenses are rarely used to provide magnifications larger than about 20×, but they can be used for 100× or more (see Section 2.1).

Although there are now many types of compound microscope they are all essentially instruments which are used:

- to visualize fine details in the structure of an object;
- to provide a magnified image of an object;
- to measure features (e.g. lengths, areas, angles) in the object;
- as an analytical tool to measure optical properties (e.g. reflectance, refractive index, phase change).

In all of the above functions there is interaction between light, which forms the image, the object itself and the eye–brain complex of the observer. Most modern compound microscopes provide magnifications between about 35× and 2000× and give an image which may be viewed directly, recorded by photography or now, commonly, received by a television camera for input into a computer image-analysis system. Stereomicroscopes are compound microscopes which provide a stereoscopic image. These

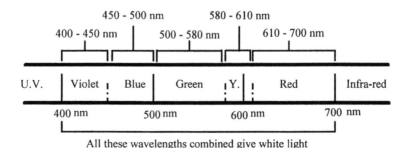

All these wavelengths combined give white light

Figure 1.2. Diagram of wavelengths of the electromagnetic spectrum. The range goes from ultraviolet (UV) to infrared (IR), and all values are in nanometres. The combination of all wavelengths from 400 to 700 nm produces light which the human eye perceives as white. The appearance of blue light is produced by wavelengths between 450 and 500 nm, green from 500 to 580 nm, and red from 610 to 700 nm.

microscopes are valuable for dissection, for manipulation of objects and study of objects where a perception of depth is useful. Stereomicroscopes usually give magnifications of between about 10× and 250×.

The radiation which is used to form the image in the light microscope is just part of the total electromagnetic spectrum of waves. The components differ in their vibration frequency (and hence in wavelength) with the higher frequencies possessing shorter wavelengths. The whole extent of radiation includes radio waves, X-rays, gamma radiation and those wavelengths from about 730–760 nm to 380–360 nm to which our eyes are sensitive and which we call 'light' (*Figure 1.2*). Conventionally in books the beam of light is expressed as a two-dimensional sine curve where the wavelength represents the colour of the light whilst the amplitude of the wave gives the information which our brain interprets as 'brightness' (*Figure 1.3*).

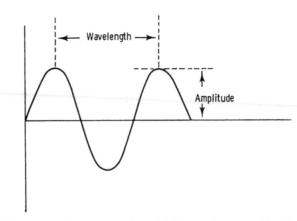

Figure 1.3. Diagrammatic representation of light as a sine wave. The distance from one peak to the next is the wavelength, usually represented by the Greek letter λ (lambda), whilst the height of the curve above the horizontal axis is the amplitude which is interpreted as brightness.

1.3   Interactions of light with matter

Light interacts with matter in several ways; an understanding of these interactions is needed in order that the user can appreciate how to control contrast and visibility of objects under the microscope. The principal interactions are:

- reflection
- transmission or absorption
- refraction
- variations of the plane of polarization
- diffraction
- excitation of fluorescence

Each of these can affect the quality and the nature of the microscopical image.

When light falls on an object, some of it is reflected; the remainder is either absorbed or transmitted.

Reflection occurs when light strikes a surface (see *Figure 1.4a* and *b*). If the surface is smooth, we have 'specular' reflection; the angle at which the beam is reflected is equal to the angle at which it arrives – in other words, the angle of reflection is equal to the angle of incidence. If the surface is rough, diffuse reflections return the light at all possible angles. Reflection is the main means by which our eyes normally receive light from objects around us. Reflection of light plays a major part in the microscopical study of the surfaces of opaque objects.

Absorption is the name given to the process whereby the amplitude of light which passes through an object (the *transmitted* light) is reduced in comparison with that which passes around it. It is one of the principal

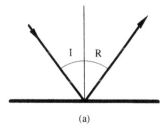
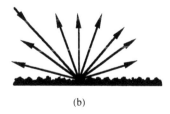

(a)   (b)

Figure 1.4. Diagrams of reflections from smooth and rough surfaces. (a) Specular reflection from a smooth surface. The angle of incidence (*I*) is equal to the angle of reflection (*R*). (b) Reflection from a rough surface occurs at all possible angles.

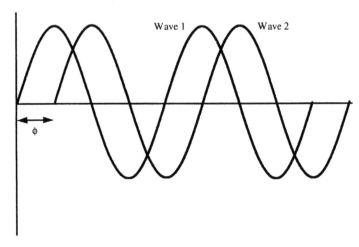

Figure 1.5. Diagram of two light waves with the same amplitude but differing in phase. Wave 2 is advanced by an amount ϕ relative to wave 1. Such a phase change is not detectable by the human eye, but it is converted by the phase-contrast microscope into an amplitude (brightness) difference, which is detectable.

interactions between light and matter used to obtain contrast in biological microscopy. As in reflected light microscopy, where absorption of some wavelengths only may affect the colour of the reflected light, in transmitted light microscopy absorption may take place either for all wavelengths (resulting in a dark object), or only over a limited range of wavelengths (so resulting in a coloured object). For example, if the wavelengths making up white light are absorbed below say 600 nm, then the object would appear red. Many objects around us absorb light in this way so that we can see them without any further treatment (e.g. a stained-glass window). Most biological cells and tissues, however, are transparent and thus are generally prepared for examination by making them absorbent with various dyes. Completely transparent objects (such as the majority of living cells or tissues) do not alter the amplitude of light passing through them, but quite often they do change the phase of the light (*Figure 1.5*).

The difference between the light which has passed through the object and that which has passed around it is called the phase change or optical path difference (ϕ in *Figure 1.5*). This phase change can be changed into an amplitude difference by allowing the waves to interfere. If two separate waves of equal amplitude and in phase with each other are combined, then the resultant wave has twice the amplitude of the original waves (*Figure 1.6*, top). If the waves are of equal amplitude but are exactly out of phase (*Figure 1.6*, bottom), then their combination will cause the resultant amplitude to be zero. This principle is used in the phase-contrast microscope.

Refraction occurs when light passes from one medium into another of different refractive index – as from air into glass. The refractive index of

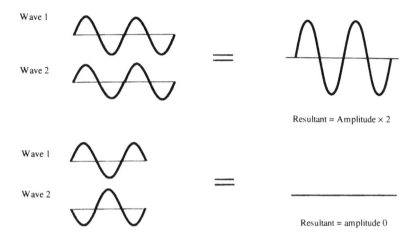

Wave 1

Wave 2

Resultant = Amplitude × 2

Wave 1

Wave 2

Resultant = amplitude 0

Figure 1.6. Diagrams to show the combination of two waves of equal amplitude but differing phases. In the upper part of the figure the waves are in phase and the resultant would have twice the amplitude of either original. The lower figure shows the effect when the phase difference is 180 degrees. The waves cancel with a resultant value of zero.

any medium is the ratio of the phase velocity of the electromagnetic waves of light in a vacuum to their phase velocity in the medium itself. Because of the different speed of light waves in the two media, the beam is deviated from a straight line (see *Figure 1.7*). The relationship between the angles of the normal to the surface and the incident beam and the refracted beam are governed by Snell's law, which states that $\sin I.n1 = \sin R.n2$, where I and R are the angles of incidence and refraction and $n1$ and $n2$ are the refractive indices (RI) of the two media (*Figure 1.7*). Refraction is fundamental for the operation of glass lenses,

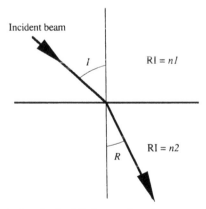

Incident beam

I

RI = $n1$

RI = $n2$

R

Figure 1.7. Bending (refraction) of light passing from one medium to another. The refractive index (RI) of the less dense medium is $n1$ and that of the more dense is $n2$. Snell's law states that $\sin I/\sin R = n2/n1$, where I is the angle between the incident ray and the normal, R the angle between the refracted ray and the normal and $n1$ and $n2$ the two refractive indices.

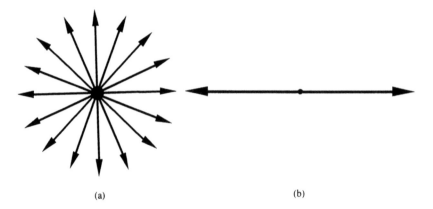

(a)   (b)

Figure 1.8. (a) A light beam considered to be emerging from the plane of the paper towards the observer has its waves orientating in all possible planes. It is unpolarized light. (b) A beam of light with its plane of vibration restricted to one direction (here the horizontal) is called plane-polarized light.

and is thus vital for microscopy. Refraction may occur between a specimen and its surroundings, or between different parts of the specimen. Where a specimen is being studied by virtue of its absorption, the differences in RI between the specimen and its surround should be minimized; alternatively, contrast in a transparent non-absorbing specimen may often be increased by maximizing refractive index differences between the specimen and its surroundings.

Polarization of light occurs in nature, but in microscopy it is induced artificially. A light beam normally contains waves of many different frequencies, with all possible phase relationships, and vibrating in all possible planes (*Figure 1.8a*). If the vibrations of the light waves are restricted to one single plane, then the light is called plane polarized (see *Figure 1.8b*). Such rays may behave differently in their passage through some types of crystalline material according to their orientation with respect to the molecular structure of the material. The speed of the rays vibrating in one direction differs from that of rays of light vibrating at right angles to this direction. Since there are two refractive indices, such materials are called *birefringent*. If the two polarized waves were initially in phase, then a phase difference between them would be generated by the time they leave the material. The polarized light microscope allows such phase differences to be used as a means of generating contrast.

Diffraction is the scattering of light that occurs when a beam passes an edge in an object; diffraction appears as a fanning-out or apparent bending of the beam, allowing the light to extend into the shadow areas (see *Figure 1.9*). The amount of such bending is related to the wavelength, being larger for longer wavelengths. Thus it is not possible to produce an absolutely sharp image of any object because the diffraction limits the

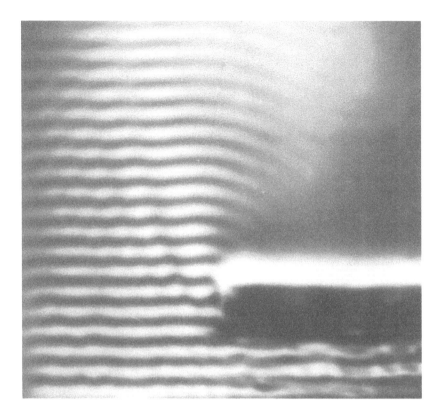

Figure 1.9. Waves in a ripple tank. This photograph shows waves on the surface of water in a ripple tank which behave in a similar manner to light waves. Plane waves at the bottom of the picture are bent (= diffracted) by an obstacle (on the lower right) into the shadow area.

resolution. It is diffraction which causes the image of a pin-hole to be not a single spot of light, but rather a central disc surrounded by a series of rings of decreasing intensity – the 'Airy disc' (see *Figures 1.10* and *5.9*). The German physicist Ernst Abbe first explained the importance of diffraction for the resolution of fine detail in the microscopical image, a subject considered later (Chapter 5).

Fluorescence is the name given to the process by which energy from light in the shorter wavelength regions of the spectrum (green, blue or ultraviolet) is absorbed by an object and then almost immediately re-emitted as light of longer wavelengths. This effect is used in everyday life in the ultraviolet devices used to test banknotes. Some objects are intrinsically fluorescent (auto-fluorescent) and thus appear self-luminous under the microscope if ultraviolet light is used. Such auto-fluorescence is often very weak, and fluorescent dyes may be added to the specimen so as to produce higher contrast images with a much brighter fluorescence when blue or green light is used for excitation of the fluorescence.

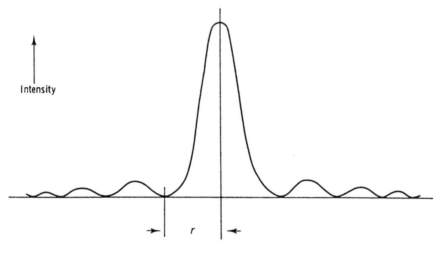

Figure 1.10. Plot of the intensity of light across the diameter of an Airy disc which represents the image of a small circular pin-hole in an opaque film. The letter r indicates the radius of the pin-hole. If the intensity of the central disc is arbitrarily represented by 1, then the intensities of the first, second and third rings, respectively, would be 0.017, 0.004 and 0.0016. This distribution of energy with 85% falling within the area circumscribed by the first dark ring would be much changed if aberrations were present.

1.4   Lens aberrations

All lenses suffer from aberrations, defects which impair their ability to produce an image which is an exact copy of the object. As far as possible all such defects are corrected in the compound lens systems used in microscope objectives. It should be remembered, however, that even if all the aberrations were to be eliminated, diffraction at the lens aperture would still cause some imperfection in the image.

There are six basic aberrations, two of which (spherical and chromatic) affect the whole field, whilst coma, astigmatism, curvature of field and distortion affect only off-axis points. In the objective lenses used in modern microscopes only spherical aberration is significant, since it is often introduced into the image by incorrect use of the microscope or by using unsatisfactory preparations. In modern objectives the rest of the aberrations have been reduced by the manufacturer to such a small extent that their effect on the image is slight. Most users need have little concern about them in practice, but as many microscopists still use older lenses all the aberrations will be described briefly.

1.4.1   Chromatic aberration

The refractive index (RI) of glass is not constant throughout the whole range of wavelengths and its variation throughout this range is called its 'dispersion'. In the visible range the RI is greater for the shorter wavelengths of violet and blue and less for the longer orange and red. There are two principal types of optical glass with differing properties. Crown glass has a low refractive index and a low dispersion whilst flint glass has a high refractive index and a high dispersion. With a lens made of a single type of glass, white light from a common axial point will not be brought to one focus; the blue rays will be focused nearer to the lens than the green and red rays (*Figure 1.11a*). This is called 'axial chromatic aberration' and its presence would result in an image appearing surrounded by fringes which vary in colour according to the focus. Axial chromatic aberration is corrected by combining lenses: a convex lens of crown glass with low dispersion is used with a concave lens of flint glass which has a higher dispersion. In theory, rays of all wavelengths could then be brought to a common focus. This is shown diagrammatically in *Figure 1.11b*. In practice, even with the very elaborate modern glasses with carefully controlled dispersions, this is only partially achieved. With the highest quality lenses (apochromatic objectives), however, the correction of chromatic aberration approaches very closely to the ideal.

Some older lenses may be well corrected for axial chromatic aberration but still show a residual defect called 'lateral chromatic aberration' or 'chromatic difference of magnification'. This is only noticeable for off-axis points because the lens may exhibit a different focal length for the peripheral rays at the extremes of wavelength; this means that the lens has unequal magnification for light of different colours. At focus the red image of an object would be larger than the image in blue (*Figure 1.11c*), so that objects would appear to be surrounded by a colour fringe. This is much more marked when the object is very contrasty and is especially noticeable with short focal length (high magnification) objectives. Chromatic difference of magnification was corrected in older microscopes by the use of a so-called 'compensating eyepiece' (see Chapter 5) which introduced an equal but opposite chromatic difference of magnification into the final image. The chromatic correction of condenser, objectives and eyepieces is determined by the manufacturer and the only improvement which can be effected by the user if chromatic aberration is present is to use monochromatic light. Modern microscope objectives are available with several different degrees of chromatic correction (see Chapter 5).

1.4.2 Spherical aberration

Spherical aberration may be defined as a defect of a lens whose surfaces form part of a sphere. The rays emanating from a point in the object on the optical axis at different angles to the axis intersect the optical axis in the image space either before (under-correction) or behind (over-correction) the ideal image point formed by the paraxial rays, that is

(a)

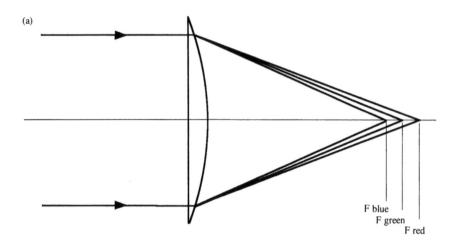

F blue
F green
F red

(b)

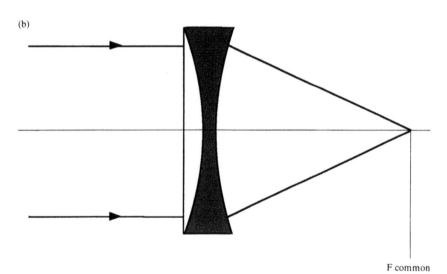

F common

(c)

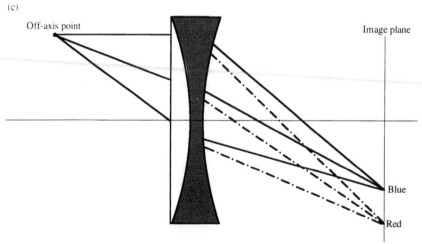

Off-axis point

Image plane

Blue

Red

Figure 1.11. (a) Diagram of axial chromatic aberration. This diagram of the path of white light through a simple plano-convex lens shows (to a greatly exaggerated degree) axial chromatic aberration. Blue rays are focused closer to the lens (at F blue) than either the green or the red rays which come to a focus at F green and F red, respectively. (b) Correction of axial chromatic aberration. The addition of a concave lens of higher dispersion glass to the plano-convex lens corrects the axial chromatic aberration. All the rays now come to the same focus at F common. (c) Lateral chromatic aberration. A diagram of a lens system corrected for axial chromatic aberration but still showing lateral chromatic aberration. The position in the image plane of an off-axis point imaged by red and blue rays is shown. The difference in position between these is the chromatic difference of magnification which would cause colour fringing in the image.

those which are parallel to the optical axis. There will be no point of sharp focus but a 'zone of confusion' along the lens axis (*Figure 1.12*).

The greater the aperture of an uncorrected lens, the more pronounced will be the spherical aberration. It can be corrected to some extent by reducing the aperture so that only the most axial rays participate in image formation. This will, however, severely limit the resolution so it is not a practicable solution.

Spherical aberration may be minimized by suitable choice of the lens surface curvatures and their orientation; in order to minimize spherical aberration in a single lens, the refraction (i.e. the deviation) of the light should be divided equally between the lens surfaces. It is less for a plano-convex lens when its flat side is towards the object than for the same lens reversed or for a biconvex lens. In practice spherical aberration is corrected by combining a positive convex lens with a negative concave element which has spherical aberration in the opposite sense (i.e. the marginal rays have their focus further away from the lens than the axial rays). By suitable choice of the radii of curvature of the positive and negative elements of an objective, the spherical aberration of the

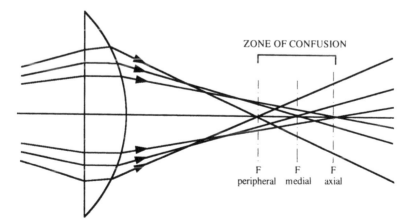

Figure 1.12. Spherical aberration in a simple lens. The axial rays are focused at a point (F axial) further from the lens than that for medial (F medial) or the peripheral rays (F peripheral). No sharp image exists and there is an extensive 'zone of confusion'.

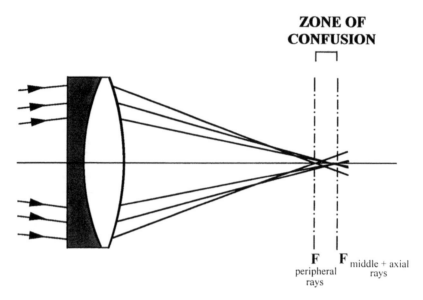

Figure 1.13. Correction of spherical aberration. Combination of a positive and a negative (concave) lens reduces the spherical aberration, and so the zone of confusion, considerably. The middle and axial rays are now focused at a common point (F middle + axial rays) and only the peripheral rays come to a focus in a different plane (F peripheral rays) nearer to the lens.

whole may be largely cancelled out for a given set of optical conditions (*Figure 1.13*). There may still, however, be some small difference of focus between the peripheral and middle and axial rays, which is known as the 'zone of confusion'.

Correction for spherical aberration may be impaired by:

- damage to the lens;
- the use of an incorrect tube length;
- the use of a coverslip of incorrect thickness (or the use of a slip if the objective is designed to work without one and vice versa) *or* an excess of mounting medium between the cover and the specimen;
- studying a detail deep in a thick specimen;
- the use of an incorrect refractive index of coverslip, slide or mounting medium.

Of these the first is beyond the control of the user and, needless to say, damage to objective lenses should always be avoided by care in their storage and handling. Very little can be done about aberration due to the second point, since on modern stands there is no means of adjusting the mechanical tube length. With the high degree of standardization of slides, coverslips and mountants now available from the suppliers, the final point is seldom a problem. This means that spherical aberration which is controllable by the user is largely due to the use of a cover of the wrong thickness or the use of excessive thickness of mountant. The latter is a very significant source of image degradation with 'dry' apochromatic objectives (i.e. those designed to operate with air between the mount and the front element of the objective; see Chapter 5). If such an objective has a numerical aperture in excess of 0.7, and numerical apertures of 0.95 are not unusual in this class of objective, then spherical aberration due to incorrect lens adjustment is often a problem. Some 'dry' objectives of high numerical aperture are fitted with a correction collar which alters the internal separation of elements in order to adjust the correction of spherical aberration. Spherical aberration is well illustrated in *Figure 1.14a* and *b* which shows the silvered opaque bars of an Abbe test-plate. When studied with a high-power objective they appear sharp when the objective is correctly adjusted (*Figure 1.14a*) but the

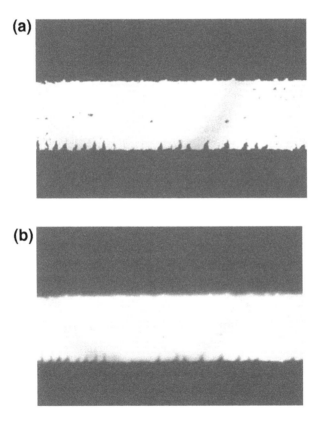

Figure 1.14. Image degradation by spherical aberration. Two photomicrographs of the same specimen (two opaque silvered bars of an Abbe test-plate) taken with a high aperture dry objective fitted with a correction collar to compensate for variations in cover thickness. Photograph (a) was taken with the correct setting for the cover in use and the image is sharp and contrasty. Photograph (b) illustrates the image degradation caused by the presence of spherical aberration introduced by deliberately setting the correction collar incorrectly. A similar loss of image sharpness would result even if the spherical aberration was corrected but the front lens of the objective was contaminated by immersion oil or grease from a finger.

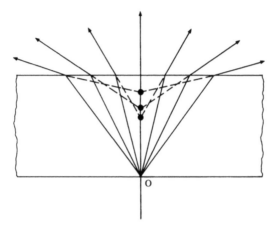

Figure 1.15. Spherical aberration introduced by coverslip thickness. A diagram to show how rays of differing obliquity, arising from an object O mounted immediately beneath a cover, appear to originate at different depths along the optical axis. This has an important consequence when using the highest quality of corrected objectives of high numerical aperture. The makers correct the spherical aberration for a cover of known thickness but using incorrect thicknesses of cover (or the presence of a thick layer of mountant between the object and the cover) will negate this correction and introduce spherical aberration into the image.

definition is lost and the image appears 'hazy' in the presence of an appreciable degree of spherical aberration (*Figure 1.14b*).

If spherical aberration is present due to a coverslip (the thin parallel-sided sheet of glass that covers the specimen, with a refractive index the same as that of the slide and the mountant) which is too thick, then, as the rays emerging from the same point in the object deviate more and more from the optical axis, they appear to emanate from a series of points at higher and higher levels in the cover (*Figure 1.15*). This gives spherical aberration, and the blurring of the image and loss of contrast illustrated in *Figure 1.14*.

It should be noted that spherical aberration due to incorrect coverslip thickness is not a problem with oil-immersion objectives, that is those in which there is optical homogeneity between the slide, specimen, coverslip and the front lens of the objective by introducing a drop of oil of the correct refractive index. If a covered specimen is examined with an objective designed to work with uncovered objects the image will again suffer from spherical aberration.

1.4.3 Coma

This is an aberration associated with points which lie off the axis and which, as a consequence of coma, are imaged as a conical or

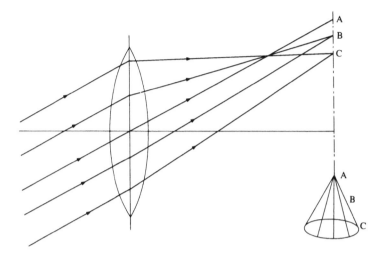

Figure 1.16. Coma in a lens. Rays from a distant off-axis point are focused at A for the axial ray, B for intermediate rays and C for peripheral rays. When the image of a point is considered for rays passing through all areas of the lens, the image would be a small cone of light with a 60 degree angle as shown at the bottom right of the figure.

comet-shaped blur. The more axial rays contribute to the image of the 'head' whilst the more oblique rays form a series of images of decreasing intensity but increasing size with their centres more and more displaced from the axis. *Figure 1.16* shows diametral rays in the plane of the paper from an off-axis point at infinity being focused by a comatic lens. The image ABC is linear; it does not lie in the plane of the paper but is tilted towards the reader at an angle of 30 degrees. The diametral rays at right angles to the original rays would give a linear image behind the paper tilted at 60 degrees to the first image. Similarly rays from all other diameters contributing to the formation of the image will be found at angles in front of and behind the ideal image plane, so that the total image of the point will appear as a small cone of light (*Figure 1.16*, lower right). Coma is a most objectionable aberration but is not found in today's objective lenses. A lens corrected for coma has equal magnification in its different zones for points in the field which are off-axis. When a lens is corrected for coma throughout, say, two-thirds of its field diameter and is nearly free of coma for the remainder, and in addition is free of spherical aberration, then it is said to be 'aplanatic'.

1.4.4 Astigmatism

This aberration often affects oblique rays passing through the lens; the effect is that the image of an off-axis point is not reproduced as such but its appearance differs according to the focal plane chosen (*Figure 1.17*). At the nearest point to the lens (B′–B′ in the figure) the image of the point appears as a horizontal line; as the focus is moved away from the

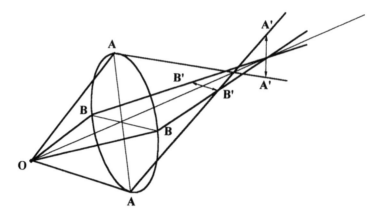

Figure 1.17. Image formation by an astigmatic lens. The rays passing through the lens diameter B–B form a linear image at B′–B′ whilst those passing through diameter A–A form a similar image at A′–A′ orientated at right angles to the first image.

lens, the image becomes an ellipse, then a circle at the point midway between the two extremes and finally it reverts to a vertical line at A′–A′. Astigmatism in lenses intended for use with light (unlike electron lenses) can be corrected only by the lens designer choosing suitable combinations of convex and concave surfaces, together with the use of glasses of different dispersion.

1.4.5 Field curvature

Although the above aberrations may be well corrected for a lens system the image, especially of achromats, could still show the defect of field curvature (*Figure 1.18*). Instead of the image lying in a plane parallel with that of the lens, it falls upon a surface which has a spherical

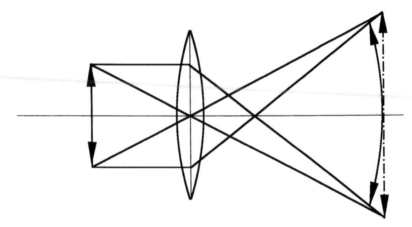

Figure 1.18. Field curvature in a lens. When this aberration is present the image of a linear object (arrows) does not lie in a plane but on the surface of a sphere.

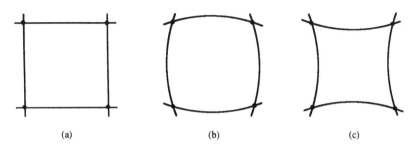

Figure 1.19. Distortion. A truly rectangular object (a) will be imaged as at (b) if 'barrel distortion' is present, and as at (c) in the presence of the converse 'pincushion distortion'.

curvature. This defect, if present, will be especially troublesome with high aperture objectives of short focal length, when only either the centre of the field of view or the peripheral area will be in sharp focus. The presence of field curvature makes it impossible to focus both simultaneously. Curvature of field may be partially corrected for medium-power objectives by the use of a special eyepiece. Specially computed 'plano' or flat-field objectives contain a large number of optical elements, often of specially formulated glasses. If one wishes to obtain an image in which all parts of a thick specimen are in focus at the same time, this may be achieved by the use of a confocal microscope, a very special technique which is considered by Sheppard and Shotton (1997).

1.4.6 Distortion

This aberration results if the magnification of the lens varies across the diameter of the field of view. Greater magnification at the centre of the field results in so-called 'barrel' distortion while the converse gives 'pincushion' or negative distortion (*Figure 1.19*). Modern objectives are so well corrected that distortion should not be noticeable.

1.5 Scales of magnification

Those starting to use the microscope usually wish to use high magnifications from the beginning. Even those well versed in using a microscope often start with a 10× objective which, with a 10× eyepiece, gives a total magnification of about 100×. This is a mistake for two reasons. Firstly, especially for beginners, a total magnification even of 10× is a large conceptual jump – equivalent to making someone twice the height of a house. This is particularly the case with children, for whom an instrument providing a total magnification of more than 10× or 20× gives an image to which they may find it difficult to relate because of

(a)

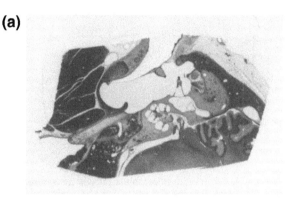

(b)

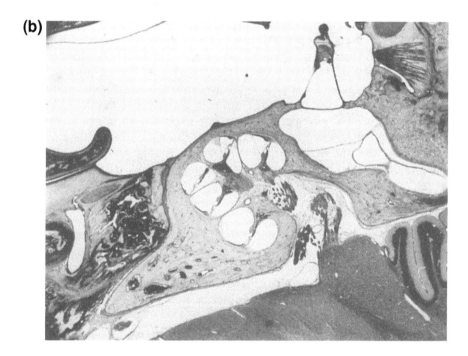

Figure 1.20. Four photographs to show progressive increase in magnification. They were made from the one original slide, without filtration, from successive negatives (Kodak Technical Pan 120 roll film, frames 56 × 72 mm); all thus had the same processing. Each was printed without dodging at the same enlargement, and the whole frame of each is shown. (a) Life size. On the negative, the slide was 25 mm wide. The preparation is of a section of cat ear. This life-size view clearly shows the interrelationships of the organs, with the auditory meatus (centre, top), the malleus attached to the eardrum below this, the stapes attached to the oval window of the inner ear, the coiled cochlea, and other structures. (b) magnified 7 times. A 1 × plan-fluorite objective was used on a compound microscope, but an alternative method would be to have used a single lens only. The detail of the stapes is seen, and the coiled cochlea clearly contains the organ of Corti. Much of the same detail could be recognized with a 10× hand lens. (c) Magnified 70 times. A 10× plan-apochromat objective was used to provide this superb view of the organ of Corti in section. The original preparation is of the highest possible quality. The outline of the illuminated field diaphragm is shown sharply

(c)

(d)

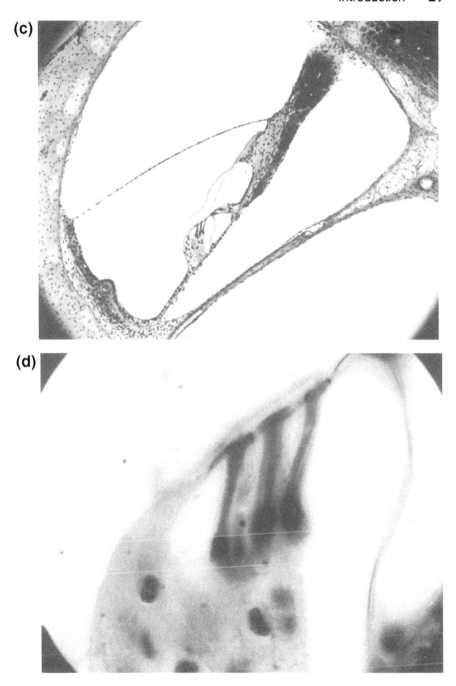

focused. (d) Magnified 700 times. A 100× plan-apochromat objective of NA 1.40 was used with an achromatic-aplanatic condenser oiled to the slide. The high quality of the image (which is actually an optical section of the preparation) allows details of the cells to be made out fully. The high quality of the condenser is proved by the sharpness of the image of the field diaphragm; this image cannot be sharp if the condenser does not have almost the same quality of corrections as the objective. The whole series is a good illustration of increasing magnification revealing more and more about less and less!

their limited experience: novice microscopists of any age are in this same position. The second reason, which applies even to more experienced workers, is that much valuable information about a specimen can be obtained using magnification of 1× (i.e. with the unaided eye) followed by 10× (using a hand lens); only then is it useful to employ the higher magnifications of the microscope. This is a very valuable point which needs emphasizing! It is well illustrated by the series of micrographs (*Figure 1.20a–d*) of a section through the mammalian ear. The relationship of, for example, the organ of Corti in the inner part of the ear (shown in the micrograph of *Figure 1.20c*, which is magnified 70×) to the remainder of the structure is completely unintelligible unless correlated with the 1× image of the whole ear presented as *Figure 1.20a*.

Progressive increase in magnification up to whatever value is needed to resolve the required detail is the only correct way to proceed if maximum information is to be extracted from the specimen.

1.6 Resolution and magnification

Most beginners place undue emphasis on magnification. This is often needed in order to make the details of our specimens easily visible. However, this is only part of the requirements; in addition to magnification a microscope image also requires *resolution* of the detail present in the original object. Although with a light microscope a specimen could quite easily be magnified by as much as 10,000×, such great enlargement would be pointless unless there was a comparable increase in the resolution. Magnification which provides no new information is known as *empty magnification*.

Resolution, or more correctly, the minimum resolved distance, is the least separation between two points at which they may be distinguished as separate. For such resolution to occur, the images of the points, as they fall on the cells of the retina, must be separated by at least one cell. The angle formed by these points at the eye, known as the viewing angle (B in *Figure 1.21*), is about one minute of arc when calculated for an object at 250 mm from the eye, the conventional nearest distance of distinct vision. This corresponds to the ability to distinguish as separate points which are about 0.07 mm apart (the points O–O in *Figure 1.21*). An object of the same size but further away (O′–O′ in *Figure 1.21*) would subtend a smaller angle (i.e. less than one minute of arc), and then the two points would not be resolved.

If detail in the object is closer together than this value, it can be resolved only if the viewing angle is increased. Either the object must be moved

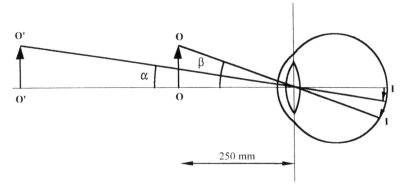

Figure 1.21. The viewing angle subtended at the eye by two points O–O at the nearest distance of distinct vision (250 mm) is angle β; if this is greater than about one minute of arc then the retinal image I–I will reveal the points as separate. If the points are more distant (O′–O′) then the visual angle α will be less than one minute of arc and the points will not register as separate because they do not fall on separate visual elements of the retina of the eye.

closer to the eye or (if it is already at the nearest distance of distinct vision) a microscope of some kind must be used. As will be explained later, the resolution of a given objective is governed by its numerical aperture (NA, which is an expression related to the maximum angle of light rays it can accept from the object plane) and the wavelength of light used. In practice, for a very good quality oil-immersion objective of NA 1.4 used with green light (which has a wavelength of c.550 nm), the resolution is about 0.25 μm. Since the practical upper limit of the NA of an objective is about 1.4, increase in resolution is possible only by the use of shorter wavelengths. Blue light with its shorter wavelength may be used but the value is limited by the fact that the sensitivity of the eye falls off dramatically as the wavelength becomes shorter. In the past ultraviolet light has been used, together with objectives and eyepieces made of quartz, in an attempt to achieve greater resolution. Nowadays for resolutions significantly better than 0.25 μm and hence the possibility of greatly increased magnifications, it is necessary to use radiation of much shorter wavelength – commonly a beam of electrons – in the electron microscope.

The theoretical resolution of a microscope system may well be greater than the actual resolving power attained in practice. If the instrument is not used properly this will certainly be the case. Resolution will suffer if the instrument is used without understanding and, at higher magnifications in particular, images showing detail that is not actually present can easily be produced.

A further characteristic of the image is that it must be visible. This seems a trite statement, but it is not resolution of fine detail alone or magnification which determines what can be seen. Sufficient *contrast* must also exist in the image for image detail to be distinguishable from surrounding material. Contrast may already exist in the specimen, it may be added to the specimen during its preparation or it may be created optically. A non-microscopical example might help here: if you are standing in a high wind, you can't see it, as it is difficult to create the contrast necessary to show air movements, but you know there is a strong wind by observing its action on things around you, as well as by the feel of it on your face.

Thus, a microscopical image if it is to be of practical use must have adequate resolution and contrast and be at a suitable magnification. The subject of contrast is covered in detail in the RMS Handbook No. 34 (Bradbury and Evennett, 1996).

1.7   Beyond this book

This book is one of the Royal Microscopical Society's series of Handbooks on microscopy. Others in the series cover more specialized topics; some of these, as well as other books which may be found useful, are listed in the Bibliography (p. 121). Many excellent books on microscopy were written 100 or more years ago, and much of what they have to say is as relevant today as when it was written. Notable examples are the books by Carpenter and Dallinger, Hogg, and Spitta listed in the Bibliography.

References

Bradbury, S. and Evennett, P.J. (1996) *Contrast Techniques in Light Microscopy.* Royal Microscopical Society Handbook No. 34. BIOS Scientific Publishers, Oxford.
Sheppard, C.J.R. and Shotton, D.M. (1997) *Confocal Laser Scanning Microscopy.* Royal Microscopical Society Handbook No. 38. BIOS Scientific Publishers, Oxford.

2 The hand lens

2.1 Hand-held magnifiers

These include the basic domestic magnifying glasses, together with higher-power, corrected hand lenses in a variety of mounts. Low-power magnifiers should not be despised for scientific purposes. For example, when scanning a large histological section before using a higher-powered compound microscope, even a quite ordinary magnifying glass is useful. Older designs of magnifier were made of glass, and thus heavy to use unsupported for very long; modern designs are moulded in plastic and are thus much lighter to hold. Some modern plastic magnifiers have aspheric surfaces which help in minimizing 'pincushion' distortion in the image. Pincushion distortion, which causes straight lines at the periphery of the image to appear bowed outwards, is a common fault in older glass magnifiers. Magnifying glasses are sold in many sizes and shapes, and one low-power example should be to hand (see *Figure 2.1*).

Higher-quality hand lenses, with magnifications from 5× to as much as 20×, are sold in several qualities and styles. Some have only a single lens, whilst other magnifiers are made up from two or even three cemented components; these compound lenses provide an achromatic image, that is one without colour fringes surrounding the edges of contrasty objects. For magnifications of about 5× a single lens will suffice but for magnifications of 10× (and certainly for 20×) an achromat gives a much sharper image. The magnifiers may be carried in folding pocket-mounts especially designed for use in the field, or in fixed mounts suited for use on a stand. A 10× achromat is a very useful second magnifier to have to hand at the microscope bench.

Historically, much higher magnifications were attained by hand lenses (simple microscopes) such as those used by Leeuwenhoek which have been widely described. His had a maximum magnification of about 250× (but more typically 100×), while a modern example made from high-refractive index material by the amateur microscopist Horace Dall had

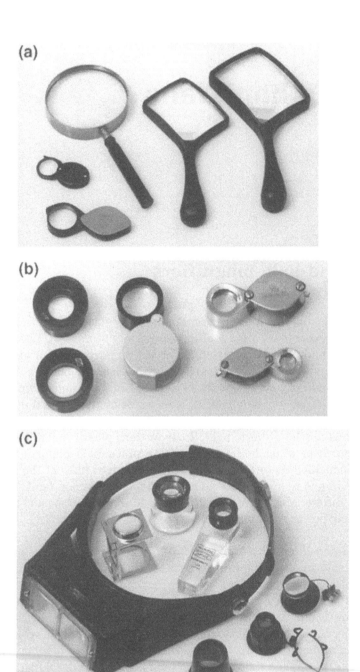

Figure 2.1. Low-power magnifying glasses and hand lenses. (a) Three magnifying glasses, of the usual pattern with a handle. That on the left has a magnification about 2×, and is made of a single glass lens of diameter 100 mm. On the right are two modern magnifying glasses, each with an aspheric lens moulded from plastic. These latter lenses do not exhibit pincushion distortion, an image defect of most magnifying glasses. The smaller is 83 × 64 mm in size and has a power of 6.5 dioptres (focal length 150 mm). To the left of these are two single lenses, each in a folding mount for the pocket. The larger provides a flat field of 4×. (b) Corrected hand lenses of higher power. The 10× lens in the folding mount in the centre of the group is a cheap version made from two cemented components. All the rest are triplets giving excellent image quality. On the

a magnification of about 450×. The image quality of such very high-power single lenses can be surprisingly good, but all such devices require the lens to be placed very close to the eye, and are tiring in use compared with compound microscopes, so that high-power simple microscopes have long since fallen into disuse.

2.2 Supported magnifiers

Smaller hand lenses can be supported in several ways, making them easier to use. Simple feet can be built on to plastic lenses of diameter as much as 100 mm, supporting them at the correct distance for viewing flat objects, including large histological preparations. The older 'linen-prover' (see *Figure 2.1*) with folding foot is also still useful, especially in larger sizes. All hand lenses, however mounted, can be supported by ordinary laboratory clamps and bosses for temporary use in dissecting, for example. For viewing solid objects there is much to be said for supporting the lens on a flexible arm long enough to keep the base well away from the work area. Magnifiers intended especially for use as 'dissecting microscopes' are easier to support and are often carried in a swivelling arm, and provided with focusing movements, a stage, substage mirror and hand-rests. A dissecting microscope is a useful instrument for viewing mounted slides as it allows easy movement of the slide carrying the specimen to and fro for scanning.

The older watchmaker's eyeglass, held by contraction of the muscles in the edges of the orbit, still has its uses in higher powers for scanning mounts and specimen manipulation. A special type of supported hand lens is that which attaches to one's ordinary spectacles. These lenses may be flipped out of the way when not required, and come in a variety

left are two by Prior, 5× and 10×, mounted for use in a dissecting microscope. On the right are Gowlland lenses of 8× (lens diameter 15 mm) and 20×, also in folding mounts. (c) The lower group are all supported lenses. For the binocular headband magnifier, three interchangeable lens pairs (each 104 mm wide) are available, with magnifications of 2.3, 2.7 and 3.5×. Within it are shown, on the left, a large linen-prover, with an opening 1 inch square, with edges marked in eighths of an inch. The 8× Agfa lupe to its right is made entirely from plastic, and is ideal for placing on a slide for a first look. The 12× achromat below them is carried on a plastic base in which it may be focused and locked. The group of three lenses illustrates (on the left) a watchmaker's eyeglass; these are available in magnifications from 2× to 10×, and once the knack of holding them in the orbit has been learned, are very convenient in use. The other two lenses are for attaching to spectacles, and both flip out of the way when not needed. The centre one in the group is by Ary and is a 10× aplanatic with a supplementary lens attached, which allows a total magnification of 18×. The basic lenses come in a range of magnifications from 2.5× to 10× and even if one does not usually wear spectacles, these lenses are so useful that it may be worth fitting one to a pair of blank frames!

of powers. They can be very helpful in initial scanning of a specimen, since the corrections of the wearer's basic spectacles are maintained. A selection of such magnifiers is shown in *Figure 2.1*.

Similarly, a binocular headband magnifier carries a converging pair of lenses and may be used over ordinary spectacles if need be. The headband magnifiers have magnifications of as much as $3.5\times$, and are useful not only in initial scans, but also in manipulating specimens, for they allow both eyes to be used and thus provide a three-dimensional image.

Older high-quality hand lenses are as good for modern use as new ones – the image quality was optimized quite early in their history; they should be coated, of course, for maximum contrast in the image. New hand lenses of high quality are amazingly expensive!

2.3   Microfilm readers

Many laboratory offices and libraries have microfilm readers to hand, and such machines (set up for microfiche) are ideal for scanning large histological preparations. Some offer a range of magnifications via a turret of high-quality lenses, with convenient support, bright and even illumination over a wide area, and the possibility of more than one person viewing at a time into the bargain (see *Figure 2.2*). Such a machine is also valuable for evaluating colour transparencies.

2.4   Illuminating the specimen

Few refinements in illumination are needed for these low-power lenses. Ambient light suffices for most uses, supplemented by a bench lamp (with large opal bulb) where necessary. If greater intensity is needed for higher magnification simple lenses an ordinary microscope lamp has ample output and can be focused to illuminate a suitable field; this type of lamp is also very suited for work with a dissecting microscope. A low angle of illumination for reflected-light work is often helpful. The shadows provided by such low-angled lighting have the effect of throwing into relief surface details which might otherwise be overlooked. If a transparent specimen of low contrast is carried on the stage of a dissecting microscope, the lighting from underneath the stage may be contrived to be at such an oblique angle that no direct light enters the lens.

Figure 2.2. A microfilm reader in use. A Zeiss reader of the 1970s (560 mm high as pic-tured) is useful to scan preparations (and colour transparencies). Here the preparation of a section of whole human brain (on a slide 4 in. × 5 in.) is shown on the carrier intended for microfiche – which is, in effect, a gliding stage. More than one person at once can see the preparation, and the turret of four Tessar lenses (28, 37.5, 50 and 70 mm focal lengths) provides a good range of magnifications to suit most gross sections.

The specimen will, however, scatter light which will emerge at all angles and so some of it will enter the mounted lens and form an image. This image will be of reversed contrast, that is the object will appear bright upon a dark background; this is called 'darkground illumination' and this technique is very useful as a means of obtaining contrast.

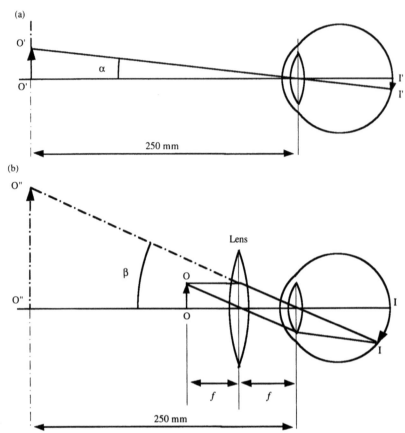

Figure 2.3. Ray diagram to show the principle of the simple microscope (hand lens). (a) Two points (O′–O′) if placed at the nearest distance of distinct vision (250 mm) subtend a small viewing angle (α) and produce a small image (I′–I′) on the retina, so that detail may not be resolved. (b) If the points (O–O) are placed instead at the focus of the convex lens with a focal length *f*, the viewing angle is effectively increased to β and the retinal image (I–I) is larger. There is an erect, virtual image (O″–O″) at infinity and detail in the object is now resolved.

2.5 Magnification

Figure 2.3 summarizes image formation with a single lens. If an object (O′–O′) is placed at the nearest distance of distinct vision, conventionally taken as 250 mm, then the object will subtend a small angle at the eye (α in *Figure 2.3a*). (This distance of 250 mm is now taken as the 'reference viewing distance' and provides the standard to which the magnification of microscopes is referred.) The image (I′–I′) formed on the retina may be so small, however, that detail in the object will not be resolved. If the same-sized object (O–O in *Figure 2.3b*) is now placed at the focus (the distance designated by *f* in the figure) of a convex lens, then the visual angle will be effectively increased to β and the retinal image (I–I) is larger. The eye sees an erect, virtual and magnified image which appears to be at infinity and detail may now be resolved. The magnification of the lens is the ratio of the two angles α and β. From *Figure 2.3a* it is clear that:

$$\tan\alpha = \frac{O'-O'}{f}$$

Whilst in the case of the magnifier (*Figure 2.3b*) we have:

$$\tan\beta = \frac{O-O}{f}$$

Since O'–O' in *Figure 2.3a* equals O–O in *Figure 2.3b*, both being the original object, we obtain:

$$\frac{\tan\alpha}{\tan\beta}$$

which is equal to:

$$\frac{250}{f}$$

that is the magnification of the lens. This may be expressed in words as the nearest distance of distinct vision divided by the focal length of the magnifier. Both distances must, of course, be specified in the same units.

In the past the power of lenses, such as those used in magnifying glasses and spectacles, were designated in dioptres, that is the recipro-cal of the focal length expressed in metres. Values for lenses described in this way range from about 1.25 to 5 dioptres. For the higher-power magnifiers currently in use, the actual magnifications are specified and the figure is often engraved on the mounting of the lens.

3 Low-power stereomicroscopes

3.1 Basic considerations

For providing lower magnifications with an erect image and full perception of depth a stereomicroscope is an invaluable instrument. It does not, of course, have to be used primarily as a stereo device, for it is also a useful intermediate-magnification instrument in its own right, providing those lower magnifications which are often unavailable on ordinary compound microscopes. Stereomicroscopes normally have a long working distance and considerable depth of field (with enhanced 3-D effect), and thus facilitate manipulation of the specimen: they are the instrument of choice when minute dissections have to be made, for example. Stereomicroscopes are generally used for the examination of objects by reflected light, although some are designed to fit on bases which carry an in-built lamp for the examination of slides or transparent objects by transmitted light. If the stereomicroscope is not supplied with a purpose-built trans-illuminating base, it is often possible to examine objects with transmitted light simply by arranging for the microscope to be supported over the object which has been placed on a light-box.

Of all the types of stereomicroscope that have been used over the years, two designs are still important. The older type, designed by Greenough and introduced by Carl Zeiss in 1897, is essentially two separate compound microscope tubes with matched objectives and eyepieces mounted side by side at an angle. The other, originated by Zeiss in 1946, has a common main objective mounted in a single body with two separate, parallel eyepiece tubes.

3.2 Greenough-type stereomicroscopes

These instruments, although introduced a hundred years ago, are still manufactured. Each separate tube contains an erecting prism above the objective, and this type of instrument is the one of choice where three-dimensional effect is important. Eyepiece separation is normally achieved by slight rotation of the prism boxes. Magnification may be altered by exchanging pairs of objectives or eyepieces, or by incorporating a magnification changer. The latter may take the form of a series of interchangeable pairs of lenses inserted into the light path, each pair forming a Galilean telescope. Alternatively, movable lenses may be inserted in the light paths to provide a 'zoom' system. These instruments (see *Figure 3.1*) generally have relatively small fields of view, and must be well treated if they are not to get out of alignment. They were often made with the ability to interchange between various bases and arms to give much versatility in use, and especially to allow them to be carried over quite large specimens to work on a particular part (see Section 3.4).

3.3 Common-main-objective microscopes

These, the most widely used nowadays, have a large-diameter objective, with two further separate lens systems mounted behind it, each carrying an erecting prism below the eyepieces which are mounted in parallel tubes. Such instruments have their primary image plane parallel to the plane of the object, and the images are simultaneously in focus across a wider field of view than in the Greenough type, where, because of the inclination of the tubes, there is a tendency for the image to be slightly out of focus towards the edge on each side of the centre line of the field of view.

It is important that such instruments have an objective of very high quality. Much of the light forming the image passes at oblique angles through the periphery of the objective, and their numerical apertures are commonly lower than might be supposed, as a double beam path has to be accommodated within the objective. Many modern instruments have a 'zoom' magnification changer built in, and are extremely easy and convenient in use.

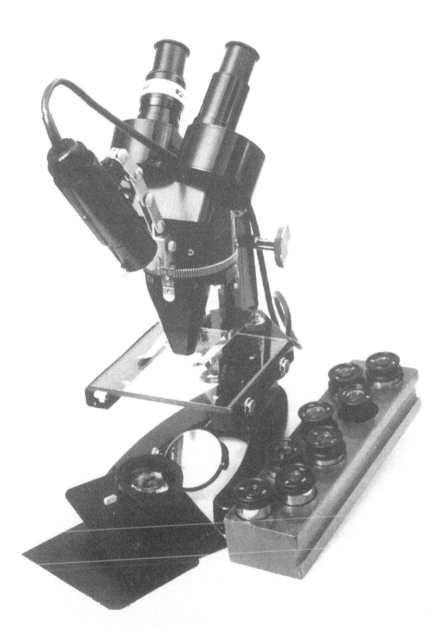

Figure 3.1. Greenough-type stereomicroscope. This M6000 stand made by Cooke, Troughton and Simms in the 1960s is on its transmitted-light base, which will tilt if need be, and which carries a double-sided mirror, large thick glass stage (easily interchanged), a substage condenser capable of illuminating a circle of diameter 30 mm, and a double-sided metal blanking-plate. Hand-rests also attach, and the instrument itself will easily detach to be mounted on a long-arm stand. The three paired objectives on the nosepiece work with five pairs of eyepieces, giving a range of magnifications from 8.75× to 200×, with fields of view from 20 to 1 mm diameter. The attachable 6 V/15 W lamp for reflected-light work is on a jointed arm which allows it to be placed anywhere over the stage. In spite of being made about 1960, this is still an excellent instrument in use.

3.4   Macroscopes

An instrument superficially similar to the stereomicroscope is the macroscope, designed principally for photomacrography. Macroscopes have only a single imaging system, with an objective of large diameter (*Figure 3.2*), often apochromatic (and very expensive). Its aperture is significantly higher than that fitted to a typical stereomicroscope. Macroscopes are usually fitted with a binocular viewing head and a camera port or, now, multiple camera ports. These camera ports usually have provision for mounting 35 mm bodies, large format bodies and a television camera for providing an image suitable for computer processing or display. It is common now for the camera exposures to be controlled by an automated system.

One macroscope from Leica (the Leica DM C) is especially intended for the low-power comparison of two very similar objects, for example in forensic studies, and so has two matched lens systems linked by a comparison bridge to one binocular eyepiece which presents a split field to the observer. This instrument has other typical features of macroscopes, for instance a very large stage and a long working distance for the objectives. The Makrozoom Leitz lens fitted to this instrument is a zoom lens with variable magnification and it will also accept auxiliary lenses fitted to its front element. These allow the magnification to range from 4× (when the working distance is 192 mm and the field of view is 50 mm) to 80× (when the working distance is 39 mm and the field of view is 2.5 mm). A macroscope should allow the use of both oblique and coaxial incident light and the base should have an aperture to allow transmitted light to be used if the specimen demands it.

3.5   Supporting the instrument

Although stereo instruments are mostly used for solid specimens, they also have uses in low-power microscopy with transmitted light. Stereomicroscope stands primarily intended for transmitted light use are mounted on a base which has an integral low-wattage bulb and a ground glass or clear plate glass insert to support the slide (*Figure 3.3*). Most of these bases illuminate a large diameter and although they provide ample intensity of illumination this is often of uneven intensity over the field of view so that they are often suited only to visual observation. A few makers provide bases which are capable of giving darkground illumination. Stereomicroscopes are very useful for initial scanning of slides at lower magnifications than are easily attainable with standard compound instruments.

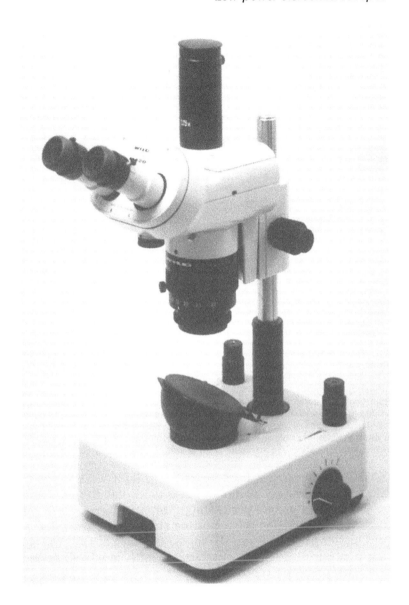

Figure 3.2. Wild M420 macroscope. This instrument from about 1980 (510 mm high as pictured) has a large lens of adequate numerical aperture to sustain the overall magnification range of 3.9× to 160×, and fields of view from 53 to 0.7 mm, all at high apertures and long working distances (still 39 mm at the highest power). These values, which are far better than those realizable with the Greenough system, are attained by using an internal 5:1 zoom mechanism, with eyepieces of 10× and 20× magnification, and supplementary lenses of 0.5 and 2×. This example of the Wild macroscope also carries a camera tube. It is shown on an illuminating base (here with a tilting cup stage added for reflected-light work) but the macroscope can very easily be fitted on a variety of other stands to allow it to be used at any angle over even large objects for reflected-light work. A vertical illuminator can be used for such work if shadowless lighting is needed, for inspecting cavities for example.

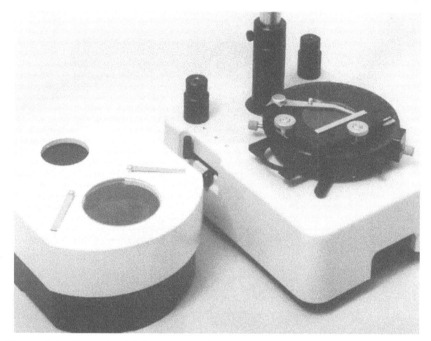

Figure 3.3. Transmitted-light bases. On the left, a Wild base which gives adequate (diffused) transmitted light and also darkground, of field diameter 25 mm. Darkground, which can be an extremely useful means of increasing contrast, is very simply set up for low magnifications with this base. On the right, a larger illuminating base (230 mm wide) for Wild macroscopes. It has three condenser lenses for visual work to match a range of magnifications and a flip-in diffuser. At the rear are bases for attachable lamphouses; an advanced polarizing stage is shown attached. This has vernier readings to the centring rotation, a push-in rotatable polarizer and mechanical movements. Similar bases of varying degrees of sophistication are available from several makers.

Stands for stereomicroscopes used for reflected-light work range from the 'lash-up' of a laboratory clamp and boss (with liability to vibration and some difficulty in precise positioning), through purpose-made ordinary stands to high-precision roller-bearing stands (see *Figure 3.4*) with consequent higher price. Perhaps the most expensive of all is the stand supporting a surgical operating microscope which may well be motor driven and microprocessor controlled.

For occasional use when high outlay is not justified, many older stands are still perfectly capable of supporting modern stereomicroscopes and macroscopes with perfect steadiness, although such stands tend to be difficult to lift on account of their weight.

3.6 Lighting the specimen

For transmitted light, the bases mentioned in Section 3.5 are suitable. Simple arrangements of a large source (an opal mains lamp, for

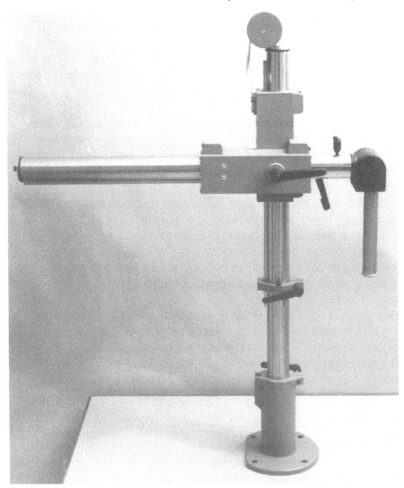

Figure 3.4. A precision stand. This Microtec precision stand (790 mm high) is bolted to the bench with a plate below for extra rigidity. The whole column rotates and clamps, and carries an adjustable collar to prevent the head from being lowered too far. The countersprung head carries a clampable arm in roller-bearings. This has a clampable carrier for the macroscope which will rotate fully and also tilt through a large angle. The whole set-up will carry a 10 kg instrument with perfect rigidity and with all adjustments free from backlash. This is an admirable example of a heavy-duty precision stand, allowing a macroscope to be positioned over large objects and set at compound angles without fear of movement during setting or in use.

example) below a pair of photographic enlarger condensers of suitable size will certainly suffice to illuminate large fields with an intensity and evenness suited to direct visual observation. For photographic purposes these are not adequate, and other measures would be needed, as described by Bracegirdle and Bradbury (1995).

For reflected light, or for less demanding direct observation by transmitted light, any lamp which can be accommodated around the stereomicroscope can be employed. It will be helpful if it can be varied in intensity and focus (so that a smaller and more intense spot can be used for higher powers), and it should be possible to vary the angle of incidence of the lighting as well as its direction. Much detail may be brought out by suitable choice of angle and direction, although these must be standardized for repetitive work.

Ordinary focusable tungsten lamps, as used with older compound microscopes, are entirely adequate for work with stereomicroscopes, especially as they can be set up at some distance from the microscope stand so as not to impede access to the specimen.

Perhaps the most versatile unit for all but the lowest magnifications is the fibre-optic lamp (see *Figure 3.5*), for this is easy to adjust in use and conveys little heat to specimens. If more than one fibre bundle is used, each with a different colour filter and set at approximately right angles to one another, then three-dimensional detail is made more easily discernible because of the two-colour shadows.

In all reflected-light work with stereomicroscopes it is imperative that various directions and angles of incident illumination be used on each specimen, so as to reveal all the surface structure present. If one standard type of lighting is used all the time, it is easy inadvertently to orientate a specimen so that significant detail may be invisible.

3.7   Image formation

The image formation in both the Greenough instrument and the common-main-objective instruments is summarized by *Figures 3.6* and *3.7*. In both types the overall magnification is the product of the objective magnification and that of the eyepiece, together with any factor due to a magnification changer or tube lens. The prisms in the ray path act to give an erect and right-way-round image.

Reference

Bracegirdle, B. and Bradbury, S. (1995) *Modern PhotoMICROgraphy.* Royal Microscopical Society Handbook No. 33. BIOS Scientific Publishers, Oxford.

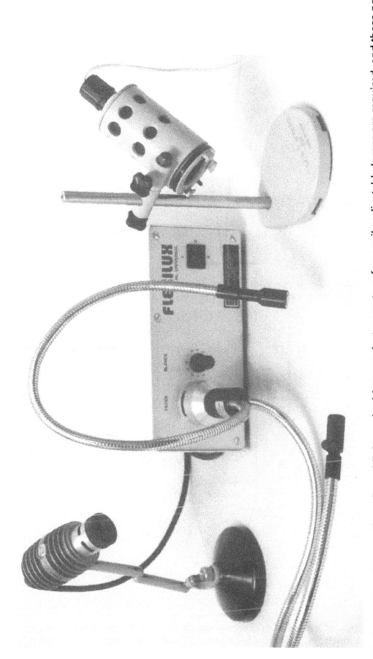

Figure 3.5. A selection of lamps. For reflected-light work with stereo instruments, a few easily adjustable lamps are required, and these need not be of the most modern designs. On the left is a small Watson 6 V/21W focusable lamp on a jointed arm, making it easy to adjust. On the right is a 1970s (but still current) Russian lamp, with 8 V/20 W prefocus bulb. Although intended as a source for a compound microscope, this works very well with low-power stereomicroscopes and, even when bought new, is inexpensive. In the centre is a 1980s lamphouse, with dual fibre-optic light-guides attached. The source is a 20 V/150 W tungsten–halogen lamp in a dichroic reflector, and the lenses at the ends of the light-guides are focusable. Their illumination is essentially free from heating effects.

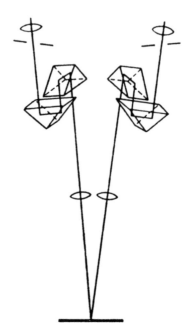

Figure 3.6. Diagram of the ray path in a Greenough-type binocular. Note that the ray paths are inclined, and each contains two prisms to produce the erect, right-way-round image.

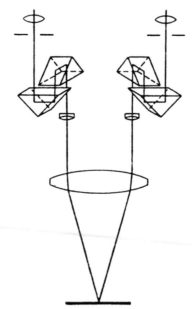

Figure 3.7. Diagram of the ray path in a common-main-objective binocular. The parallel ray paths (here much simplified) for the right and left eyes are clearly shown, passing through the periphery of the large main objective. The small lenses just below the erecting prisms are the tube lenses.

4 Compound microscopes

4.1 The compound microscope

A compound microscope is one in which a real magnified image produced by one lens (or lens system) called the objective is further magnified by another – the eyepiece – which typically forms a final, magnified, virtual image for observation by the user's eye. This applies also, of course, to the stereomicroscopes and macroscopes described in the previous chapter, but when the term 'compound microscope' is used, it generally means one not specially adapted to produce stereo images. The production of a real microscope image for projection or photography will not be considered in this book; interested readers should consult Bracegirdle and Bradbury (1995).

For the image produced by the compound microscope to be useful, three main criteria must be fulfilled. Firstly, the lens systems must reveal the required detail, that is the image must possess adequate *resolution*. This is largely controlled by suitable choice of the microscope objective, although incorrect setting of the instrument will degrade the performance of even the highest quality objective. Secondly, the image must also have sufficient *contrast* to allow the eye or recording medium to distinguish what has been resolved. Contrast may often be provided by suitable preparation of the object under study, but contrast enhancement may also be achieved by the use of suitable techniques of microscopy and correct usage of the instrument. Thirdly, the total *magnification* must be sufficient so that the image of separate points resolved by the microscope falls upon separate cones in the retina of the observer's eye.

Microscopy carried out by biologists usually involves objects which are sufficiently thin and transparent to allow light to pass through them. Alternatively, biologists often kill their material, embed it in some supporting medium such as paraffin wax or a resin and cut very thin slices (called sections) of it. These, after colouring with dyes and mounting, are also studied by transmitted light. Transmitted-light microscopes have a stage for supporting the specimen, which has a central hole to allow the light to pass through, and the illuminating apparatus (mirror or lamp-house and condenser) is mounted below the stage. Much microscopy, however, especially in the industrial field and in metallurgy, ceramics and materials science, requires the study of opaque objects. Microscopes for such work (see *Figure 4.8*) use reflected light. The illumination system produces Köhler illumination (see later) in a similar manner to that for transmitted light. The important difference, however, from a transmitted light stand is that the objective acts as its own condenser. Such a microscope may often also include polarization apparatus to minimize unwanted reflections from some highly reflecting specimens, and also be fitted with special objectives allowing the illumination to be arranged for darkground. Some of these instruments are also inverted in design. Haynes (1984) summarizes their use in materials science.

4.2   Microscopes with separate lamps

Student-type instruments of today, and the better-quality older ones, are not to be despised for almost any kind of microscopy. Many stands built in the 1950s (as well as some made even earlier) were very well built and, if they have been adequately looked after, will still be fully capable of good work. They should nowadays be fitted with modern objectives which have their lens elements coated so as to provide an image of high contrast, free from stray light (often called glare); thus equipped, the principal drawback of early stands is their relative lack of convenience as compared with models having built-in light sources (see *Figure 4.1*).

The positive advantage of a good monocular or binocular compound microscope, used with a separate high-intensity lamp, is that once the user understands the correct procedure for its use it can easily and cheaply be adapted for most kinds of work. Such a microscope can often accept older lenses and accessories from more than one maker. It is, of course, a nuisance to have to realign the lamp each time the instrument is reused after it has been put away, but two approaches to this problem are suggested. First, given sufficient bench space, the microscope and lamp need not be replaced in its case after every use. The lamp and stand can each be covered with a plastic bag and left in alignment; the purpose of the cabinet is for transporting the instrument and not to contain it after every use. Second, if it is inconvenient or there is too little space to allow the microscope and lamp to be left on the bench, the resetting after storage may be made much quicker if the microscope stand and lamp are mounted on a board which has stops to hold both in alignment.

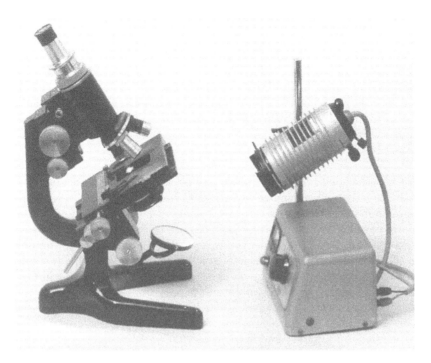

Figure 4.1. Compound microscope with separate lamp. This 1950s Watson Bactil stand (375 mm high as pictured) is set up with a 1960s Wild 6 V/30 W tungsten filament lamp in a good-quality lamphouse on its own transformer base. The drawtube has a clamping ring to allow a camera to be attached without the tube sinking under the weight. Such an arrangement is less convenient than more modern approaches, but has the merit of possessing great flexibility.

When setting up a microscope there is no virtue in using a more complex arrangement than the work in hand requires! For low-power investigation of pond animals, for example, the 5× objective can be used with a 10× eyepiece (to give an overall magnification of 50×) using an opal lamp in a shield as the source of light, the concave side of the mirror and no condenser. For the higher magnifications a lamp fitted with a collector lens, properly aligned with the plane mirror and a substage condenser, will be required to provide Köhler illumination (see Section 6.4).

The flat side of the mirror is used to reflect light into the condenser, which focuses a circular spot of light on to the specimen, through it into the objective and on through the eyepiece to the eye or camera. The diameter of the area of the specimen which is illuminated is governed by the setting of the illuminated field diaphragm, together with the lamp distance and the condenser focal length. The objective is the key element in the microscope system and, although some mineralogical stands and a few older stands for high-power photomicrographic work have provision for centring the objective, usually all the other optical elements must be centred to it. The aim is to place the lamp filament, the lamp collector lens and its iris (the illuminated field diaphragm), the mirror, the condenser iris (the illuminating aperture diaphragm), the condenser, the objective, the eyepiece, and the eye, all on the one optical axis (Section

7.1 provides detailed instructions on how to do this). These adjustments sound formidable, but are done in practice almost quicker than it takes to write about them. Most of these adjustments may be done once and for all in stands with built-in light sources. As suggested above, some of these adjustments for microscopes used with separate lamps may be minimized by mounting them on a board together with the lamp which is thus held in a fixed position relative to the microscope.

Older microscopes might be monocular or may have a binocular head (see *Figure 4.2*). This type of head is not designed to give stereoscopic images, but it is more restful to use both eyes if an instrument is used for long periods. Two matching eyepieces are then required, and adjustments may be needed for both the interocular distance and to compensate for possibly different characteristics of the two eyes. This is a once and for all operation for any given user and the binocular head will have provision for making these adjustments. Stands from the 1960s and later often have a trinocular head, which has two tubes for the eyes and, at the same time, one for a camera (*Figure 4.3*). If photomicrography is to be carried out such a head is convenient.

Older monocular microscopes when used with a separate lamp were often tilted back from the vertical to provide greater comfort in viewing. Better-quality instruments were totally stable in this position, even when inclined to the horizontal (as for photomicrography or when using a micro-aquarium on the stage). For use with fresh liquid preparations, the microscope has to be kept vertical, of course. In such a case, it may be helpful to place the stand on a lower bench than usual, to allow for more comfortable viewing. As may be seen from *Figure 4.2*, the binocular eyepiece tubes fitted to older microscopes may be inclined so as to provide a comfortable viewing position without the need to incline the limb. Inexpensive student microscopes available today are also fitted with an inclined eyepiece tube and now few possess an inclinable limb.

4.3   Microscopes with built-in light sources

Modern microscopes have their light sources incorporated into the stand (*Figure 4.4*), making them very convenient to use; such stands are not inclinable but have inclined eyepiece tubes. Earlier microscopes had mains or low-voltage tungsten-filament light sources and the illumination was generally not of the Köhler type. While these light sources were entirely suited for most visual observations, the use of phase contrast at high powers might produce only a dim image, and for projected images, as in photomicrography, such sources are certainly too feeble. Many student-level stands with this type of illumination are currently still being supplied by Russia and China.

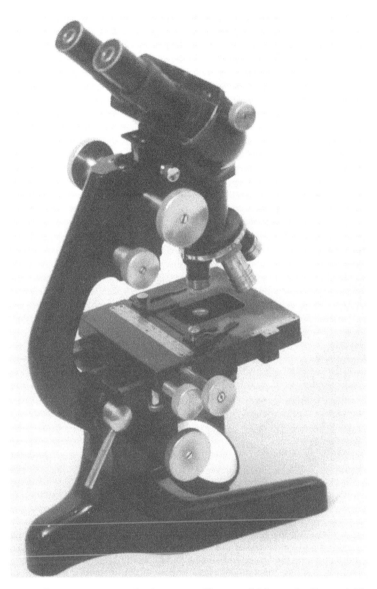

Figure 4.2. A 1950s compound microscope. The stand (shown in *Figure 4.1* in mono-cular mode) is easily fitted with a binocular head, as the body tube is divided specifically to allow this. Such older stands were usually very solidly made and provided with adjustments to compensate for wear. Thus, although this stand has had full use for many years, it still works sweetly and has all that is needed for high-quality work. Such instru-ments need not be replaced for the latest model unless more convenience in use is an imperative requirement. Note that this is quite the reverse situation in comparison to some current equipment (e.g. computers).

Later and more sophisticated microscopes had tungsten–halogen sources and Köhler illumination. By the 1970s microscope lamps gave light of an intensity as high as anyone could need. Bulbs powered by 12 V and rated at 100 W were usual and these were generally mounted

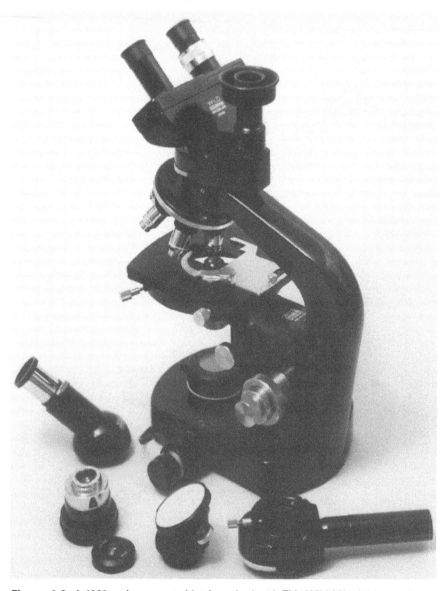

Figure 4.3. A 1960s microscope with trinocular head. This Wild M20 (390 mm high as pictured) is a good early example of the modular approach to compound microscopes. The basic stand contains the lamp, and all else can be assembled as needed. This example has a centring rotating mechanical stage and sextuple nosepiece with high-quality objectives. There is a tube to provide a camera outlet while still allowing the binocular head to be used. In front is an inclined monocular drawtube head which will fit in place of the binocular, a very high-quality achromatic-aplanatic condenser giving 1.3 NA if required, a two-sided mirror with precision adjustment which fits instead of the built-in lighting if special sources are required, and a horizontal camera tube for adding a closed-circuit television camera (CCTV). Such instruments are still perfectly capable of the highest-quality work.

in an outside lamphouse to allow the heat to dissipate (*Figure 4.5*). Such instruments tend to be rather larger than their counterparts of 20 years before, and often are made of lighter metals. Modern design and machining methods produce microscopes which operate very smoothly, and modern objectives, often constructed from new optical materials and

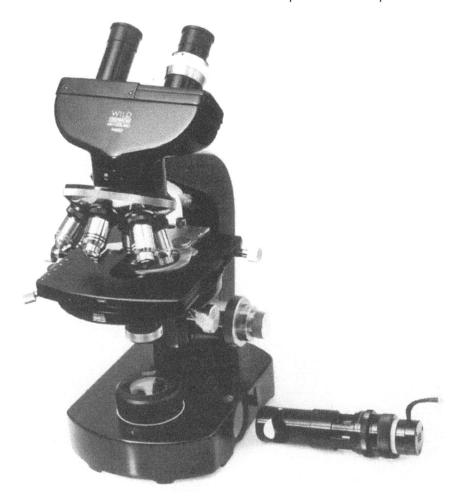

Figure 4.4. Microscope fitted with an integral tungsten filament lamp. The M20 microscope as seen in *Figure 4.3* is shown here with its lampholder withdrawn, to reveal the 6 V/20 W lamp (in special cap) in its centring insert; the power supply was external. The substage condenser lacked centring screws; setting up Köhler illumination was carried out by moving the illuminated field diaphragm itself.

designed and made with help from computers, now provide superb image quality.

4.4 Integrated microscopes

The most advanced instruments from the 1970s and later provided integral automatic camera facilities (35 mm and/or 4 × 5 in. and/or CCTV), interchangeable modules to provide a range of illumination possibilities, and often are fitted with more than one light source (*Figure 4.6*). With

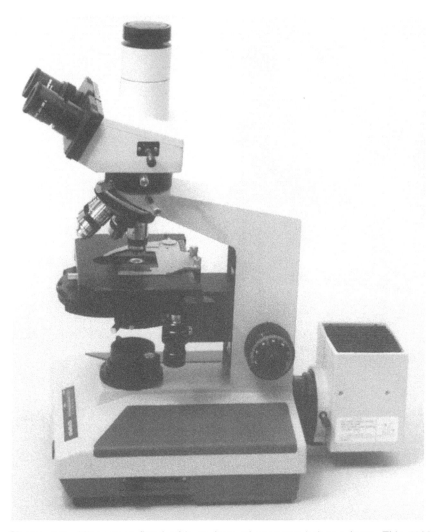

Figure 4.5. Microscope fitted with an integral tungsten–halogen lamp. This 1980s Olympus BHS instrument (505 mm high as pictured) has the power supply in its base and the 12 V/100 W lamp is in a housing at the rear. This particular stand has a wide variety of optical parts in its outfit, and is here fitted with a trinocular head. This represents the developed form of the compound microscope, perfectly suited for all kinds of work by using various interchangeable parts.

such microscopes, the use of infinity-corrected objectives (see Section 5.1) is important as it allows accessory optical components to be placed in the light path immediately behind the objective without upsetting correction for spherical aberration. Such accessory components cannot easily be used with a microscope fitted with the older type of objectives which are corrected for use with a fixed mechanical tube length.

The microscope designs of the 1970s and 80s remain very suitable for general microscopy, as well as for more advanced work which might need a variety of illumination techniques. These microscopes are well made and convenient in use, but one note of caution is needed. Their

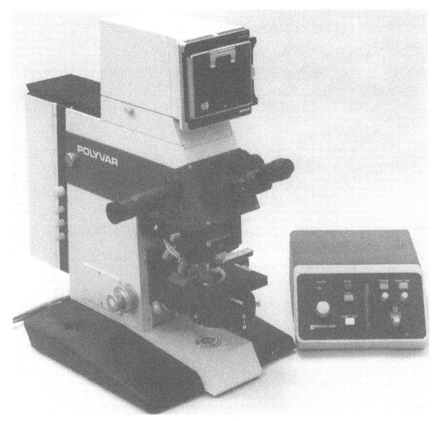

Figure 4.6. Integrated stand with dual light sources. This Reichert Polyvar of about 1981 was among the most advanced of its day. The lamphousing at the rear contains a tungsten–halogen lamp and also a xenon arc. Either could be used for transmitted and/or reflected light, and both could be used together at will. At the far side is a drawing tube and on the left-hand side of the body there is a CCTV outlet. A 4 × 5 in. camera (which could be replaced by a 35 mm back) is mounted on the top of the body. The block above the nosepiece houses various accessories, including those for phase contrast, polarized light, differential interference contrast, reflected light, transmitted light and darkground, all of which are fully interchangeable. On the right, the control box for the photographic functions allowed the operator to vary the parameters so as to provide a fully controlled exposure. The hand-rests at each side allowed the operator to work for long periods without undue fatigue, and all the mechanical adjustments worked with extreme smoothness and precision.

electronics (used to control the automatic cameras) are now outdated, and if they go wrong it might be difficult and expensive to have them put right.

4.5 Automated microscopes

The latest, largest stands now offered for light microscopy are fully automated. This means that a touch on a button selects a magnification, and then the objective is changed by a motor as is the substage condenser and the image projection system; all is optimized quite automatically for

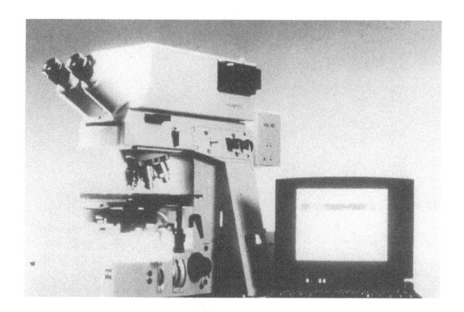

Figure 4.7. Fully automated microscope. The Axioplan 2 'robot microscope' has dual light sources for reflected-light and/or transmitted-light illumination and is equipped with automatic motorized condensers, objectives, stage, focus, fluorescence filter changers and output selection. Köhler illumination is routinely set up for every power of objective (plan-fluorite or planapo, magnifications from 1× to 100×) automatically, and a scale bar is included in each frame (35 mm to 4 × 5 in. and CCTV). A notebook computer contains the wide-ranging controlling software; the chosen settings are remembered and displayed and all can be controlled remotely if need be. The optics are infinity-corrected with all varieties of contrast available, and a wide range of accessories is provided. This is a current 'state-of-the-art' light microscope, versions of which are offered by several manufacturers. (Photograph from a colour transparency kindly supplied by Carl Zeiss Ltd.)

quality of illumination, resolution, and even focus. Changes in type of illumination are also fully automated, as is use of the range of built-in image recording devices (*Figure 4.7*). All of this has to be set up initially and be optimized for individual preferences, but once this is done it is all remembered for the future on the built-in computer (which may also perform image storage and manipulation).

Such stands are intended for intense use in the laboratory where several different microscopical techniques might be required in quick succession and when results must be obtained rapidly. For this purpose they are ideal and here may well justify their high price. Because of this high price (including their servicing costs) such microscopes are not in widespread use.

Figure 4.8. Microscope arranged for use with reflected light. This Olympus BHS micro-scope needs no substage for it is used with reflected light only. The head can be carried on a variety of stands, the objectives are corrected specifically for uncovered specimens, and the lamphouse contains a 12 V/50 W tungsten–halogen lamp with associated con-densers and irises to provide Köhler illumination. A polarizer is carried in the main head, which will accept a variety of viewing and recording heads as well as Nomarski prisms to match each objective and provide differential interference contrast. The simple base on which it is carried in the figure is 160 mm wide.

4.6   Specialized microscopes

A few specialized types of microscope are sometimes encountered. One is the petrological microscope, with built-in accessories for investigating crystals and other materials using polarized light. Robinson and Bradbury (1992) provide further details, and this type of instrument will not be further described here. As mentioned in Section 4.1, many instruments are intended solely for use on opaque objects using reflected light. These are often called 'metallurgical microscopes' (*Figure 4.8*).

In biological tissue-culture laboratories 'inverted' microscopes are often found. In an inverted microscope the lamp and condenser are mounted *above* the stage. The condenser of this type of microscope is usually designed to bring the light to a focus at a greater distance from its top lens than would be the case with a typical transmitted-light instrument. This is to allow room on the stage for a thick culture chamber contain-ing living tissue and good access to the (usually fresh) specimen so that it can be manipulated. Much ancillary equipment can be supported round the stand. The objectives are fitted on to the body tube which is here *below* the stage, and a prism takes the image-forming rays back up to an inclined eyepiece tube in front. Earlier models of such stands were quite simple and inexpensive, but fully automated inverted instruments fitted for fluorescence techniques, for phase-contrast microscopy and for differential interference contrast (DIC) are now quite as complicated and expensive as their upright counterparts.

References

Bracegirdle, B. and Bradbury, S. (1995) *Modern PhotoMICROgraphy.* Royal Microscopical Society Handbook No. 33. BIOS Scientific Publishers, Oxford.

Haynes, R. (1984) *Optical Microscopy of Materials.* International Textbook Company (Blackie), Glasgow.

Robinson, P.C. and Bradbury, S. (1992) *Qualitative Polarized-Light Microscopy.* Royal Microscopical Society Handbook No. 9. Oxford Science Publications, Oxford.

5 Basic concepts: objectives and eyepieces

5.1 More on image formation

When used in a compound microscope the objective forms a real magnified image of the specimen, an image which is then further magnified by the eyepiece. More important than the magnification, however, is the fact that the objective provides the resolution of fine detail. Although in practice the objective is composed of several lenses (see Section 5.4), it may, for simplicity, be considered as a single plano-convex lens. It will be remembered from Section 2.5 that when such a lens is used as a simple microscope, the object is placed *at* the focus of the lens and the image is magnified, erect and virtual. When a convex lens, however, is used with a specimen located just *outside* its primary focus, the image produced by the lens will be real, magnified and inverted (see *Figure 5.1*).

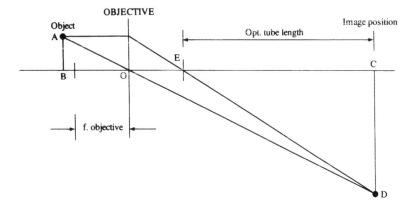

Figure 5.1. Formation of a real image by a lens. A diagram of the rays forming the real primary image by an objective. The focal length of the objective is indicated by the distance f. objective The primary image (CD) of the object (AB) is real and inverted. ABO and CDO are similar triangles used to determine image magnification.

In this figure the object is labelled AB, the focal length of the objective is indicated (Foc.objective) and the inverted real image is shown at CD. E in the figure indicates the location of the focal point which lies in the back focal plane of the objective and indicates the point where rays entering the objective parallel to the optic axis cross the optic axis on the other side of the lens. Other rays, however, arise by diffraction of the light at the object (see Section 5.5) and all these diffracted rays come to a focus also at point E. The real image from the objective is called the primary image and is located at CD in *Figure 5.1*.

In the microscope this primary image is arranged to be located at the focus of the eyepiece (F ep in *Figure 5.2*). For simplicity the eyepiece itself may be regarded as a single convex lens which thus acts like a simple microscope giving more magnification and a virtual image (GH) which is focused on the retina of the eye by the cornea and lens of the eye itself. If a real image for a camera is needed then a further lens is used above the eyepiece to achieve this. All the bundles of rays leaving the eyepiece intersect at the exit pupil of the eyepiece or 'eyepoint'. This is located at the eye lens at a point labelled EP in *Figure 5.2*; this is a distance equal to F ep. The exit pupil was formerly called the 'Ramsden circle', and is where the pupil of the observer's eye should be placed in order to see the whole of the field of view. If the eye is above the exit pupil, then the peripheral rays cannot enter the eye and so the diameter of the observed field is reduced. As the magnification of the eyepiece is increased, so the exit pupil becomes located closer to the upper surface

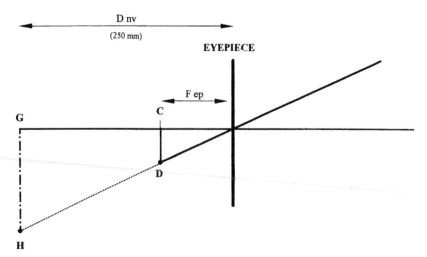

Figure 5.2. Formation of a virtual image by the eyepiece. The eyepiece is drawn as a single lens with a focal length denoted by F ep. The primary image formed by the objective lens is located at CD, whilst GH is the final virtual image at D nv, the nearest distance of distinct vision (taken as the reference viewing distance). At this reference viewing distance of 250 mm, a primary image of size CD would appear to have the size GH.

of the eye lens which makes the eyepiece difficult to use, especially for spectacle wearers.

The final virtual image has the same orientation as the real primary image although both are inverted with respect to the specimen itself. In most theoretical treatments this image (GH in *Figure 5.2*) is often regarded as being located at infinity, to be viewed by a totally unaccommodated, relaxed eye. In practice users often adjust the microscope so that the final image is located at the nearest distance of distinct vision (D nv in *Figure 5.2*) which is conventionally taken as 250 mm.

The total magnification of the microscope may be calculated by multiplying the nominal objective magnification by the eyepiece magnification. If precise measurements are to be made, then it is best to determine the exact magnification by calibrating the system. This is done using an accurate calibrated scale called a stage micrometer; full details of how this is done are given in Bradbury (1991).

The distance between the back focal plane of an objective and its primary image plane is called the *optical tube length*; this is illustrated in *Figure 5.3* (which shows a conventional compound microscope). It is worth noting that the diaphragm in the eyepiece is placed in the plane of the primary image. In this figure an internal diaphragm eyepiece (see Section 5.9) is drawn and here the first lens of the eyepiece (the field lens) is located before the diaphragm. This lens will affect the position of the primary image and hence the apparent optical tube length. This caveat does not apply to external diaphragm eyepieces. The *mechanical tube length* is set by convention. In the nineteenth century this was 250 mm; until recently 160 mm (or 170 mm for some makers) was standard for

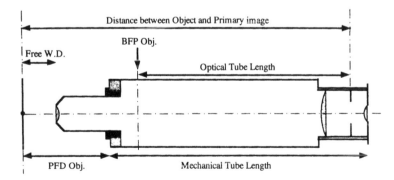

Figure 5.3. Diagram of the various dimensions of a microscope used with an objective corrected for a finite tube length. The mechanical tube length extends from the shoulder of the objective to the eyepiece rim, whilst the optical tube length is the distance from the back focal plane of the objective (BFP Obj.) to the primary image plane. The free working distance of the objective (Free W.D.) is shown and the parfocalizing distance of the objective is indicated (PFD Obj.).

microscopes fitted with conventional optics, but with the introduction of infinity-corrected objectives (see Section 5.4 below) on modern microscopes the concept of mechanical tube length has become irrelevant.

Figure 5.3 illustrates this mechanical tube length for a conventional microscope and also indicates some other common dimensions often quoted in the literature. The distance between the objective shoulder and the specimen is termed the *parfocalizing distance* of the objective. This is now set at 45 mm since the trend for the addition of extra optical elements in the objective (for the higher powers or to give extra optical corrections) requires a longer barrel. If all the objectives in a series have the same parfocalizing distance then changing from one power to another is possible by rotating the nosepiece without any risk of damaging the specimen or the objectives.

5.2   Conjugate planes

Conjugate planes are those planes in an optical system which are equivalent: an object placed in one plane will be sharply imaged in each subsequent plane of that series. In the microscope (used with Köhler illumination; see Chapter 6, p.88), two separate series of conjugate planes may be recognized.

'*Aperture*' *Series*	'*Field*' *Series*
Lamp filament	
	Lamp iris diaphragm (illuminated field diaphragm)
Front focal plane of condenser (illuminating aperture iris)	
	Specimen plane
Back focal plane of objective	
	Primary image plane
Exit pupil of the eyepiece (Ramsden disc)	
	Retina of the eye

A knowledge of these planes, illustrated diagrammatically in *Figure 5.4* is essential in order to allow the user to understand the correct adjustment of the illumination or to be able to insert phase-plates or graticules in the correct places. Note that the two series of planes are completely separate, so that the image of the lamp filament, for example, although formed in the subsequent planes *of its own series*, never appears in any plane of the other series.

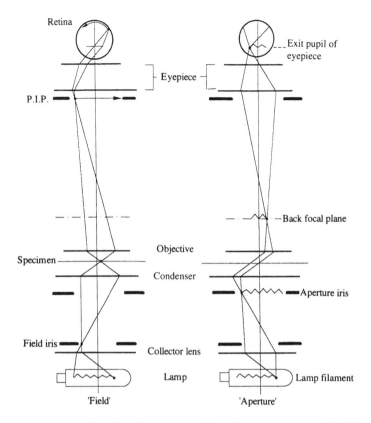

Figure 5.4. The two series of conjugate planes in a microscope arranged for Köhler illumination. On the left are shown the 'field' or 'image-forming' set and those of the 'aperture' or 'illuminating' set are illustrated on the right hand side of the diagram. P.I.P. is the
primary image plane, here shown as located just below the eyepiece. In this diagram the
illuminating aperture iris is referred to as the 'aperture iris' and the illuminated field
diaphragm is the 'field iris'.

Since measurement graticules must be in sharp focus at the same time
as the image of the specimen, it is clear that they must be inserted into
one of the planes of the 'field' series listed above. Although in practice
the primary image plane is chosen for convenience, it would be possible
to place a graticule in the plane of the illuminated field diaphragm,
when it would still be in focus at the same time as the specimen. The
image of the graticule, however, would suffer unless the corrections of
the condenser were of a high order.

5.3   Infinity-corrected systems

In the discussion of the traditional compound microscope given above,
the light beams leaving each point in the focused object are converged by

the objective and produce a real image in a primary image plane at a given distance behind the objective. The objective is thus said to work at a finite tube length. Some older monocular microscopes still in use are provided with a calibrated sliding tube (the drawtube) carrying the eyepiece, so that the actual mechanical tube length may be altered from that of the standard 160 mm. Alteration of the tube length in use was done to restore correction of the spherical aberration if this was upset by, say, an over-thick layer of mountant or a cover of the wrong thickness. The effect of spherical aberration on the image is illustrated in *Figure 1.14*. The introduction of extra optical components (such as beam-splitters or barrier filters) behind the objective of a microscope using objectives corrected for a finite tube length would lead to a shift of the primary image along the optical axis. Since spherical aberration, flatness of field and coma (see Section 1.4) are corrected for a given finite primary image distance, any displacement of the position of this image may increase such aberrations.

Most of the advanced microscopes made since *c.*1980 have a different optical arrangement which confers both mechanical and optical advantages. Their objectives are designed to have an image plane at infinity and are thus called 'infinity-corrected' objectives. In such a system the object is placed in the front focal plane of the objective and thus rays leaving the objective from any point on the object form parallel beams (*Figure 5.5*). These parallel rays are focused by a further lens in the microscope tube (called the tube lens) to form the real primary image, which is then viewed and enlarged further by the eyepiece, just as in a microscope using conventional objectives. In effect the tube lens and the eyepiece act together as a telescope, whilst the objective and the tube lens may be regarded as a 'compound objective'. A detailed treatment of

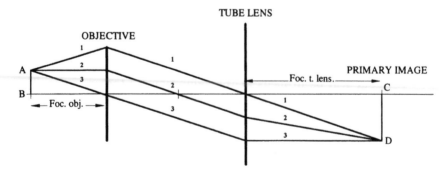

Figure 5.5. Formation of the real image in an infinity-corrected system. Three rays of light from an object (AB) imaged by an infinity-corrected objective with a focal length Foc. obj. are shown. Note that between the objective and the tube lens the light rays are parallel. This latter lens (with a focal length of Foc. t. lens) converges these rays to form an inverted real image at CD. Various optical accessories may be placed between the objective and the tube lens without significantly affecting image quality.

the optics of the infinity-corrected microscope will be found in Volume 1 of Pluta (1988).

An infinity-corrected objective cannot be said to have an intrinsic magnification of its own; the finite magnification engraved on the objective is established only in combination with the correct focal length of tube lens (often called the reference focal length) for which that objective has been designed. This has been set at different values (between 160 and 250 mm) by various makers in recent years, but the actual value for any given microscope need not concern the user. With conventional finite tube length objectives the eyepiece often had to compensate for considerable residual errors, especially lateral chromatic aberration (see Section 1.4.1), so that the objectives and eyepieces had to be carefully matched. In infinity-corrected systems it is now possible to produce with the combination of objective and tube lens, working together, a primary image which is almost free of residual aberrations. This means that the eyepieces no longer need to correct for such aberrations.

There is also a further mechanical advantage when infinity-corrected objectives are used; in a microscope fitted with a conventional objective, focusing of the image is done by moving either the stage (plus specimen) relative to the objective or by moving the entire tube with objectives and eyepiece relative to the stage. In an infinity-corrected system, however, the distance between the objective and the tube lens may be varied considerably without detriment to the image and so focusing may be done by moving only the objective. This is important since it allows a considerable simplification of the mechanical design and increases the versatility of the microscope.

5.4 Diffraction and resolution

Light is affected by its passage through a specimen; for example, some may be absorbed either partially or entirely, and diffraction also occurs. This is a scattering of light which occurs when the beam encounters an edge or irregularity in an object. It is a fanning out or apparent bending of the light into the dark, shadow areas (see *Figure 1.9*). The degree to which this occurs is related to the wavelength, being greater for the longer wavelengths. This means that it is not possible to obtain an absolutely sharp image because diffraction will be a limiting factor. As a result of diffraction the image of a pin-hole is not a sharp spot of light but has a central point surrounded by a series of rings of decreasing intensity – the so-called 'Airy disc' (see *Figures 1.10* and *5.9*). It was Abbe in 1872 who realized the importance of diffraction as a limiting factor in the resolution of fine detail by the microscope.

If we consider a narrow beam of light passing through an object with regular, periodic structure some light will pass on undisturbed. This is called the direct light or 'zero-order beam'. The detail in the object will cause the formation of a series of diffracted waves which will interfere with one another and thus be reinforced or attenuated to form a series of maxima and minima. These may be seen in the back focal plane of the objective on either side of the central beam and may be demonstrated with a regular grating as the object and a narrow axial beam of rays; in this case the zero-order and the succeeding order maxima do not overlap and are clearly visible. If a large illuminating cone is used, the diffracted orders are still present in the back focal plane but they all overlap and cannot be seen.

If we consider a small source of light illuminating two slits (A and B in *Figure 5.6*) these act as coherent secondary sources emitting spherical wavefronts (Huygens' wavelets) which can then interfere. When the crest of a wave from slit A arrives at a point at the same time as a crest originating from slit B, reinforcement occurs and a maximum is formed. Such a maximum represents the sum of the amplitudes of the two interfering wavelets. Where the crest of a wavelet from one slit arrives at a point at the same time as a trough originating from the other slit, then there is a minimum. *Figure 5.6* shows that all points equidistant from A and B produce interference maxima with no path difference. This is the zero order. At the first-, second- and subsequent-order maxima, the path length difference between the two interfering waves arriving from A and B is one, two, three, etc. wavelengths. This reinforcement and cancellation of waves is well shown by waves in a ripple tank (*Figure 5.7*) where a double dipper simulates the two slits in a grating.

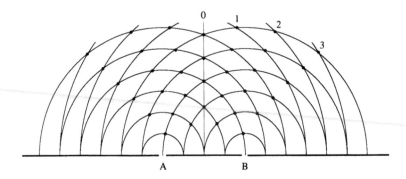

Figure 5.6. Interference arising at two parallel slits. This diagram shows the interference wavelets (Huygens' wavelets) arising from light passing through two parallel slits (A and B) illuminated by light from a single source. Points equidistant from A and B are indicated by zero where there is no path difference. The maxima resulting from reinforcement where there are path differences of one, two and three wavelengths are indicated.

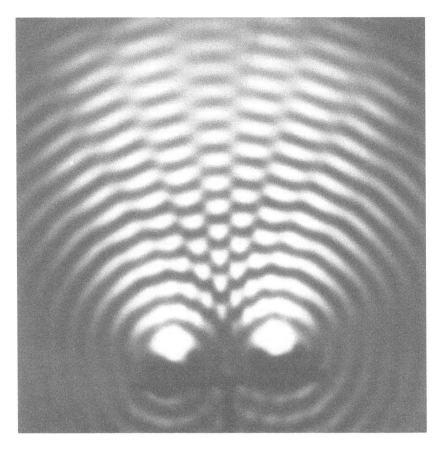

Figure 5.7. Interference between waves in a ripple tank. The wavefronts from a double dipper in a ripple tank illustrate the phenomenon of reinforcement shown in *Figure 5.6*.

Abbe studied the problem of microscopic resolution of a grating and showed that for this to occur the zero-order and at least the first-order diffracted beam must enter the objective lens. His way of thinking about resolution is often called the 'Abbe approach' or the 'diffraction at the specimen' approach. It is shown in *Figure 5.8* where the slits of the grating at X and X' are separated by the distance r. The slits are illuminated by a parallel beam of light which is either from a distant source or from a microscope condenser with its aperture diaphragm almost closed; this light may thus be assumed to be coherent. The angle by which the first-order beam is diffracted is indicated by i and the path difference between the zero-order direct light and the first-order is, therefore, the distance X'C, which is given by $r \sin i$. Since this path difference gives first-order constructive interference it must equal 1λ, and hence:

$$r = \frac{\lambda}{\sin i}$$

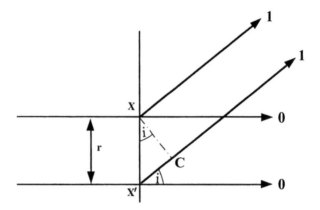

Figure 5.8. Angular separation of the interference maxima from two slits. This diagram shows the angular separation (*i*) of the interference maxima from waves arising from two slits X and X' separated by a distance *r*. The distance X'C represents a path difference of one wavelength.

This assumes the slits are viewed in air but if a medium of another refractive index (say *n*) is used between the grating and the objective lens, as in oil immersion, then the resolution becomes:

$$r = \frac{\lambda}{n \sin i}$$

Since $n \sin i$ is the numerical aperture of the objective, the final equation for the resolution may be written as:

$$r = \frac{\lambda}{NA}$$

In the equations above, the Airy factor (1.22) which is often inserted as a multiplier of the numerator λ, to take into account the circular symmetry of the objective lens, is omitted. For maximal resolution in practice, the condenser is opened fully and a cone of light with a half-angle of *i* is admitted so that the path difference is now 2× (*r* sin *i*) and the minimum resolved distance is halved:

$$r = \frac{\lambda}{2 \sin i} \rightarrow \frac{\lambda}{2 n \sin i} \rightarrow \frac{\lambda}{2NA}$$

This argument assumes that the slits are parallel and very narrow. For a constant slit separation the diffraction maxima will be closer together with blue light than with the longer wavelength of red. For a constant wavelength of light it is the separation of the slits that controls the separation of the diffraction maxima. As the spacing between slits increases the diffraction maxima become closer together; conversely, if the slits are very closely spaced (corresponding to an object with very fine detail) the diffraction maxima become further apart. This means that for some

(a)

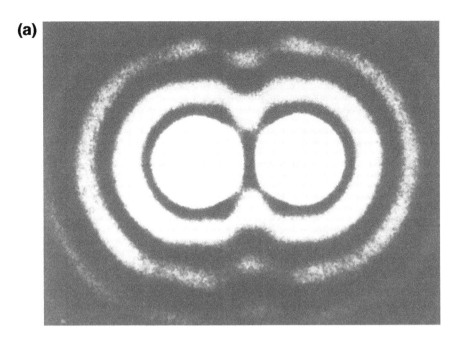

(b)

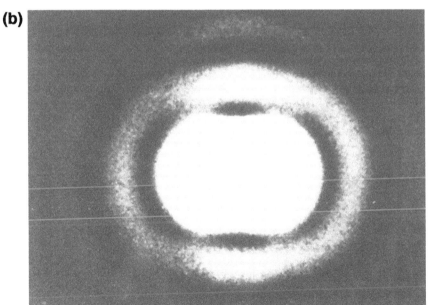

Figure 5.9. The Airy discs of two points. (a) Two separate Airy discs are distinguishable when the points are resolved as separate. (b) If the points are separated by less than the resolvable distance then the Airy discs merge, as here, into a single entity.

objectives with a limited numerical aperture the first diffraction maxima may not be accepted by the lens and the image detail would not be resolved.

As mentioned in Chapter 1 and shown in *Figure 1.10*, when we see an Airy disc from a circular pin-hole, the centre of the first interference minimum lies at a distance (r) from the centre of the zero-order maximum equal to an optical path length of 1.21 wavelengths. Two completely separate pin-holes would obviously produce two completely independent and separate diffraction patterns (*Figure 5.9a*). If the object points are brought closer together, then at some stage the Airy disc patterns overlap so much (*Figure 5.9b*) that it is not possible to say whether we are looking at one or two objects. The arbitrary conclusion, established by Lord Rayleigh, for determining whether there is resolution in the case of overlapping patterns is that resolution is said to occur when the first dark ring of one pattern coincides with the centre of the bright disc formed by the direct light, or zero order, of the second. At this point a densitometer trace would show a pattern similar to that shown in *Figure 5.10*, where there is a marked central trough in which the amplitude is about 85% of the peak value. The minimal separation in the object plane (i.e. the resolution) again may be shown to be:

$$r = \frac{\lambda}{n \sin i}$$

where λ is the wavelength, n the refractive index of the medium between the front lens of the objective and the object and i is half the angle of the entire cone of rays accepted by the lens (the half-angle of acceptance).

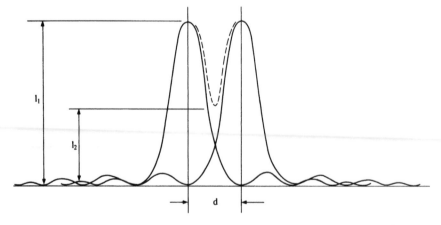

Figure 5.10. Plot of the light intensity across the Airy discs of two points. This trace represents the intensity of the light across the Airy disc pattern of two points separated by a distance d just satisfying the Rayleigh criteria (and which hence are just resolved). Note that the centre of one disc falls on the radius of the first dark ring of the second. I_1 represents the intensity of each central maximum and I_2 the intensity of the resultant, denoted by the dashed line, which has a value of about 85% of I_1.

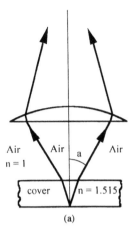 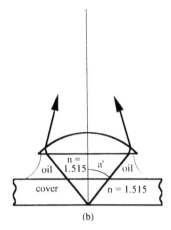

Figure 5.11. Diagrams of rays entering 'dry' and 'immersion' objectives. (a) The path of rays from an object entering a 'dry' objective after passing through a coverslip of refractive index $n = 1.515$. The half-angle of acceptance of the lens is angle a. For such an objective the numerical aperture (NA) is equal to sin a. (b) When immersion oil is added between the front lens of the objective and the cover, the value of the half-angle of acceptance of the extreme rays is increased. The NA of this lens is now given by the expression n sin a', where n is the RI of the immersion medium.

An objective can only gather a limited cone of rays from the specimen (*Figure 5.11a*). This light-gathering power of the objective will affect both the brightness of the image and its resolution. The denominator in the above expression (n sin i) is called the *numerical aperture* (NA) and it allows the performance of objectives to be compared. When there is air between the object and the front lens of the objective, n is then equal to 1 so that the NA becomes equal to sin i, the half-angle of acceptance. The maximum value sin i may achieve in practice is about 0.95. If oil of the same refractive index as the glass is added between the specimen and the front of the objective (*Figure 5.11b*) the maximum NA will be increased to about 1.4 with consequent increase in the maximum resolving power and image brightness. Other advantages of immersion lenses will be mentioned in Section 5.7.2.

The ray paths when a narrow beam of light is illuminating a grating (only three slits are drawn), together with the appearance of the back focal plane of the objective, are shown in *Figure 5.12*. In addition to the axial zero order, the objective has accepted the first two maxima on either side of the axis so the object would be well resolved.

As mentioned above, if the grating lines are very close together then the angular separation between the diffraction maxima increases. If the spacing of the grating is very fine it is possible that even for a lens of high aperture the first-order maximum would not fall within the acceptance angle. In this case, with only the central zero-order direct light present (*Figure 5.13b*), no detail would be resolved in the image. Again,

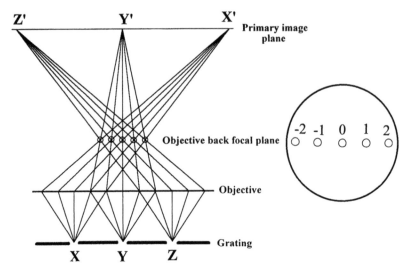

Figure 5.12. Ray paths when an axial beam of light illuminates a grating. Three slits (X, Y, Z) of an opaque grating form images X', Y' and Z' in the primary image plane. The circle on the right of the diagram illustrates the appearance of the back focal plane of the objective when the illuminating cone of light from the condenser is markedly reduced (so that the orders of diffracted light do not overlap and, therefore, may be seen easily). The direct light appears axially at 0 while the first- and second-order diffracted light is shown on either side of the axis at positions indicated by –2, –1 and 1, 2, respectively.

if the numerical aperture of the objective is reduced by closing the iris diaphragm of the condenser, the diffracted maxima might be excluded (*Figure 5.13a*) and image detail will not be seen. If the detail is to be resolved then we must either use another objective of a higher NA or the axial beam must pass through the grating at an angle to the optical axis,

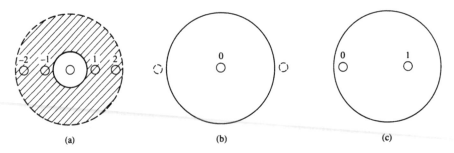

Figure 5.13. Appearance of the back focal plane of an objective under different conditions. A grating is imaged by a small cone of light. At (a) on the left, the objective aperture has been made very small by the introduction of a diaphragm so that only the direct light (the zero order, 0) enters the objective. There will be no resolution of the grating detail. If the grating's lines are very close the separation of the diffracted orders is correspondingly wide and with axial light even the first order might not enter the objective. This is shown in the centre at (b). Diagram (c) shows that if the beam of light is very oblique in a single azimuth, the zero order now enters on one side of the objective and this allows the first order to be accepted opposite to it (on the right labelled 1); there will be resolution of detail but the image quality will not be good.

that is we must use 'oblique light'. Now the zero order will appear at one side of the back focal plane and the first-order maximum can be accepted at the opposite side (*Figure 5.13c*). This would alow resolution of the detail but the image quality would not be good, since image fidelity improves according to increase in the number of diffracted orders taking part in image formation.

When the microscope is used correctly the condenser aperture is opened so that a solid cone of light which fills about 90% of the area of the back focal plane of the objective is used. This is equivalent to using oblique light from all directions at once. The diffraction maxima in the back focal plane overlap and can no longer be distinguished as separate. The resolving power is often expressed by the formula:

$$r = \frac{\lambda}{NA \, obj. + NA \, cond.}$$

so long as the numerical aperture of the objective is equal to or greater than that of the condenser. Assuming that this figure is 1.3 for each (both objective and condenser linked to the slide with oil) and that the wavelength is 0.5 µm, then d (the minimum separable distance between two points) can be calculated to be approximately 0.2 µm, a figure in close agreement with the value which would be obtained from Rayleigh's equation applied to the Airy disc.

From the above it is clear that in order to achieve maximal resolution:

- the objective should have high NA so as to accept more of the higher orders of diffracted light;
- the NA of the illuminating system must match that of the objective;
- the shortest possible wavelength of light (consistent with the maintenance of the quality of the lens corrections and the sensitivity of the eye) should be used.

In view of the reduced sensitivity of the eye in the violet and blue regions of the spectrum, in practice blue-green or green light affords the best compromise between visibility and maximal resolution. It should also be remembered that not only must the detail of the image be resolved, it is also necessary to ensure that the detail is visible. This requires *contrast* which, if it is to be achieved optically, may require breaking the second of the above points by markedly reducing the NA of the illuminating system.

5.5 The objective

The objective is the most important single optical component of the microscope. It:

- collects the light from the specimen and (possibly in collaboration with a tube lens) forms a primary real image as free from aberrations as possible for examination with the eyepiece;
- provides the resolution needed to see fine detail in the object;
- provides the greater part of the magnification needed in the image.

The construction of a satisfactory objective requires the maker to consider very carefully the choice of optical materials to be used and the design of the components, and to ensure that they are constructed and assembled to the highest standards. By comparison with 100 years ago the makers today have a much greater range of optical glasses and other materials, of specified refractive indices and dispersions, and of uniform composition, quality and stability. Both design methods and manufacturing of the lens elements have benefited from the introduction of computer-aided systems. In consequence, modern objectives have performances both in the visible and ultraviolet ranges of the spectrum which would have been unthinkable when Ernst Abbe first produced his apochromatic objectives in 1886.

Design of a microscope objective is complicated by the fact that, unlike the majority of lenses in other optical instruments, it has to have a very large numerical aperture and at the same time has a small field of view. Numerical apertures for dry objectives range from 0.05 for low-power lenses with a magnification of $1.6\times$ up to a maximum of 0.95 for dry objectives with a magnification of about $60\times$; immersion objectives may have a NA of up to 1.4 and give a magnification of between 40 and $100\times$.

Typically a microscope nosepiece would have four objectives arranged to cover the magnification and resolution range most suited to the current application. A typical set might be: $5\times$ $10\times$ $40\times$ $100\times$ (oil immersion). The chosen objectives must be suitable for use with the various contrast techniques to be used; the requirements of objectives to be used with polarized light, differential interference and fluorescence applications are especially demanding.

It is usual for the objectives of a set to be *parfocal*. This means that when one objective of the set mounted on the nosepiece is focused on the object then, when it is exchanged for any of the remainder in that set, no significant readjustment of the focus of the microscope will be needed to restore maximal sharpness to the image. When old objectives were mixed on a nosepiece this often was not true and occasionally there was danger of hitting the slide with an objective if the nosepiece was rotated without care. The lens flange to object plane distance is now standardized for conventional objectives at 45 mm (the parfocalizing distance of the objective) (see *Figure 5.3*) so this is not a problem with modern objectives. It does, however, impose design problems for objectives with a focal length longer than say 50 mm (i.e. of magnification about $4\times$) and for those high-magnification objectives with large numerical apertures which have large free working distances.

5.6 Classification of objectives

Microscope objectives may be classified for convenience in several different ways, for example those intended for use with a finite tube length and those which are infinity-corrected. Within each of these categories the basic optical corrections of the objective, especially those for spherical and chromatic aberration, separate them further. Another system of classification depends on the fact that some objectives are intended for use 'dry' (i.e. with air between the specimen and its front element), whilst others are designed for use with an immersion fluid such as water, glycerol or oil.

5.6.1 Classification according to chromatic corrections

This categorization usually divides objectives into either:

- achromats
- semi-apochromats (formerly often called 'fluorites')
- apochromats

Achromats are corrected for axial chromatic aberration for two wavelengths, one in the red and one in the blue; this means that light of these wavelengths is brought to a common focal point along the optical axis. For an achromat a plot of focal distance against wavelength would have the shape of a parabola (*Figure 5.14*). There will, therefore, be a fringe of colour around contrasty objects at a certain plane of focus. If the focus is chosen to be in the green region of the spectrum there will be a magenta halo (called the 'residual colour') around such objects. This is seldom troublesome and if achromats are used with green light, for which they are spherically corrected, they give very acceptable results for visual work and for photography in black and white. They are relatively cheap and for this reason are much used in microscopes intended for teaching and routine screening work. Many such objectives are now calculated to have a flat field and are therefore called 'plan-achromats' or have the abbreviation 'PL' in their name.

Although the residual colour cannot be completely removed by combining elements constructed of crown and flint glass alone, a great improvement may be obtained in the corrections by the use of other elements made of materials such as fluorspar; nowadays newer, specially formulated, optical glasses would be used. Such objectives were formerly

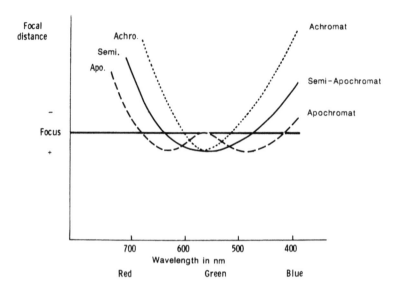

Figure 5.14. Diagrammatic plot of wavelength against focal distance for three types of objective. The plot of focus against wavelength for an achromat is a narrow parabola, whilst that for a semi-apochromat is much flatter. The curve for a very highly colour-corrected apochromat has a 'blip' in the middle as a result of bringing the foci for red, blue and green to the same level. Such an objective would show almost no residual colour, whereas the image from an achromat, if focused on a contrasty black object, has a fringe of magenta around it.

called 'fluorites' but are now more usually termed 'semi-apochromats'. They have improved colour corrections as compared with the achromats, as illustrated by the fact that the parabola in *Figure 5.14* showing the focal distance against wavelength is much flatter. This superiority of colour correction enables them to be made with higher numerical apertures than achromats of equal magnification, resulting in brighter images which show a higher resolution of detail than that of an achromat of the same magnification. Because of their superior colour corrections, semi-apochromats generally give images with a higher degree of contrast than those from an achromat.

Corrections of the highest quality are found in the 'apochromatic' objectives. These have axial chromatic aberration almost completely eliminated, as three wavelengths (in the red, green and blue regions of the spectrum) have been brought to a common focus (see *Figure 5.14*). In addition their spherical aberration is corrected for two colours rather than for only one. The numerical aperture of such objectives may be even larger than that of semi-apochromats for any given magnification. When used correctly, on a suitable specimen, apochromatic objectives produce an image of the highest possible quality in terms of resolution

Table 5.1. Typical parameters of some objectives from various makers

	Magnification	Numerical aperture
Achromats		
	10×	0.25
	20×	0.40
	40×	0.65
	60×	0.80
	100× oil	1.25
Plan semi-apochromats		
	4×	0.13
	6.3×	0.20
	10×	0.30
	16×	0.40
	20×	0.50
	40×	0.70
	63×	0.90
	63× water	1.20
	63× oil	1.25
	100× oil	1.30
Plan apochromats		
	4×	0.16
	10×	0.40
	20×	0.70
	40×	0.95
	63× oil	1.40
	100× oil, iris	0.5–1.35
	100× oil	1.40

and contrast. They are ideal for photomicrography of stained material in colour. Such corrections require a considerable increase in the complexity of the construction and thus in relative cost. An apochromatic objective of NA 0.75 could well be three times as expensive as an achromat of NA 0.65 and comparable magnification. *Table 5.1* presents some typical data for various objectives of different classes.

The older apochromats had very short free working distances and all suffered from marked curvature of field. They also suffered from another very marked aberration called chromatic difference of magnification (see Section 1.4.1 and *Figure 1.11*) in their primary image which had to be corrected by the use of specially computed 'compensating eyepieces'. These are no longer needed with infinity-corrected objectives since this aberration is corrected by the tube lens.

The pronounced field curvature of the semi-apochromatic and apochromatic objectives is the most significant of the residual aberrations, and for many years it was accepted as inevitable. When a microscope objective of this type is used, constant refocusing is carried out all the time and any point in the image may be made sharp at will. The widespread use of photomicrography and television image capture, however, makes field curvature unacceptable. As most makers now provide 'plan' objectives for all categories of correction, such objectives are now almost universally used on microscopes intended for research purposes.

5.6.2   *Dry lenses and immersion lenses*

Many objectives are designed to be used 'dry', that is with air between the front lens of the objective and the coverslip on the specimen (see *Figure 5.11a*). As mentioned above, high magnification apochromatic objectives of this type have a maximum NA approaching 1 (in practice 0.95). Such high apertures make this type of dry objective difficult to use, as their free working distance is excessively short and their correction for spherical aberration is very sensitive. The spherical aberration which may be introduced by incorrect adjustment accounts for the total inability of many users to obtain really sharp images with dry apochromatic objectives of high aperture (see *Figure 1.14*). Some dry apochromatic objectives of nominal NA 0.95 are fitted with a graduated, rotatable collar (a correction collar) which allows the elimination of much of the residual spherical aberration. Rotation of the collar alters the position of some of the elements inside the objective and allows a skilled microscopist to compensate for some of the spherical errors introduced by variable thickness of mountant or coverglass. With objectives designed for epi-illumination and uncovered specimens, large aperture 'dry' objectives have no need for correction collars since they are always focused on the *surface* of a specimen.

Immersion objectives are designed to be used with a medium other than air between the coverslip and the front lens of the objective. This allows the numerical aperture to be increased significantly, so increasing the theoretical resolving power of the objective. In addition, if the immersion medium is oil of the same refractive index (RI) as glass then possible image degradation due to spherical aberration from an incorrect cover thickness or excess of mountant will be avoided. At the same time glare due to reflections of the most oblique rays from the front lens of the objective will be avoided, so increasing the contrast of the image.

The typical immersion objective is designed to use an oil of the same RI (1.515) and dispersion as glass but objectives using water or glycerol as immersion media are also available. Indeed it is possible to obtain special immersion objectives fitted with correction collars which allow them to be used with two or three different media in turn. Glycerol-immersion objectives were originally intended for use with ultraviolet and fluorescence microscopy since glycerol does not itself fluoresce whereas the older oils did. Modern immersion oils, however, are non-fluorescent so that today oil-immersion objectives may be used for fluorescence microscopy. Water-immersion objectives are useful when temporary mounts in aqueous media are being observed since, because of the low viscosity of water, movements of the objective during focusing will not tend to disturb the coverslip.

Immersion was formerly confined to the highest-power objectives, but immersion objectives are now available with much lower magnifications

(16×, 20×, 25×, 40× and 63× from one maker, for example). These lenses allow much larger fields to be examined and of course have much higher apertures than the dry objectives of comparable magnifications. The larger fields of view are advantageous in areas such as haematology but the special value of the larger NA is in fluorescence microscopy since these objectives give images of high brightness. For any given magnification, the brightness is related to the NA by the expression:

$$\text{Brightness} = \frac{(\text{NA})^2}{(\text{magnification})^2}$$

It thus follows that for an objective magnification of 40×, the brightness given by the use of an immersion plan semi-apochromatic objective of NA 1.0 compared to that of a dry plan achromatic objective (NA 0.65) with the same magnification will be in the ratio of the squares of the two numerical apertures, that is approximately 2.4:1. Brightness is also inversely proportional to the square of the magnification, and hence low magnifications give bright images. Objectives sold specifically for fluorescence microscopy thus have large apertures and low magnifications and are often designed for water, glycerol or oil immersion. With the small amount of fluorescence often shown by specimens, such an increase in the image brightness due to the use of a high NA and low magnification is valuable.

5.7 Special types of objective

For many years biological microscopy was largely confined to the examination of fixed and stained material; materials scientists similarly used the instrument as a tool for the straightforward visual examination of their specimens. The development of objectives was thus concentrated on the design of lenses which would give maximum resolution. The development of special techniques such as the use of phase contrast, polarized light and fluorescence or the need to examine objects which could not be mounted in conventional ways led to objective design being diversified.

Phase contrast is an optical method of introducing contrast into images of specimens which, being almost transparent, are almost invisible by conventional light microscopy. Such specimens, although they have little effect on the *absorption* of the light passing through their various regions, have in fact caused some changes in the *phase relationships* between the rays of light as they leave the specimen. The phase-contrast technique invented by Frits Zernike allows alterations in phase (to which are our eyes are insensitive) to be converted into amplitude or brightness changes that we can see. In order to obtain phase contrast a phase-changing plate with a circular ring or groove worked into it is fitted into the back focal plane of a conventional objective. The theory of phase contrast and the construction and operation of a phase-changing plate is detailed in Bradbury and Evennett (1996).

For use with polarized light, it is essential to ensure that the objective itself does not contain any optical elements which are birefringent or exhibit strains which might interfere with accurate measurements of the degree of birefringence of the specimen.

Special long-working-distance objectives are also available. Often these are conventional lenses but sometimes mirror elements have been used, working in combination with lenses, to increase the working distance for any given focal length. Long-working-distance objectives are of value for the examination (through the wall of the flask) of living material in culture, for use with hot-stage microscopes when melting-point determinations are being carried out or when micromanipulation requires the insertion of probes and other instruments between the specimen and the front lens of the objectives.

Although not part of the objective, magnification changers may be mentioned at this point. These are lens systems fitted into the tube which, when brought into the optical path, allow a variable increase in the magnification of the primary image. They are much used in low-power stereomicroscopes. Many microscopes intended for use with polarized light have an extra focusable lens (the Bertrand lens); when brought into the optical path this acts in conjunction with the eyepiece to form a small telescope to give a magnified view of the back focal plane of the objective. This is intended for viewing interference figures produced by specimens viewed in polarized light with an analyser in the crossed position (i.e. with its plane of maximal transmission at right angles to that of the polarizer). In addition a Bertrand lens provides a useful means of setting the working aperture of the condenser to suit that of the objective in use.

Table 5.2 summarizes the markings often found on the barrels of objectives.

5.8   The eyepiece

The eyepiece is used:

- as a magnifier to view the real primary image formed by the objective so that the detail which it has resolved may be seen;
- in some cases to complete the correction of residual aberrations in the primary image;

Table 5.2. Markings on the objective barrel[a]

1. Words for type of objective – letters are in black for standard, red for polarized light and green for phase-contrast objectives:
   PLAN (or PL) – flat field
   FL, FLUOTAR, NEOFLUAR – fluorite
   APO – apochromat
   No marking – achromat
   Type of immersion medium:
   W or WI – water
   Oil (or oel) or HI – oil
   Glyc – glycerol
   No marking – 'dry' lens
   Magnification
   Numerical aperture
   Other features:
   LWD – long working distance
   Korr – adjustable coverslip correction

2. Figures for mechanical tube length and cover thickness for which objective is corrected (e.g. 160/0.17):
   ∞   infinity-corrected
   160   standard 160 mm tube
   –   not critical
   0.17   standard coverslip 0.17 mm thick
   0   uncovered

3. Coloured rings to indicate magnification:
Black 1.25×	Brown 2.5×, 3.2×	Red 4×, 5×
Orange 6.3×	Yellow 10×	Green 16×, 20×, 25×, 32×
Blue 40×, 50×	Dark blue 63×	White 100×, 150×, 200×

4. Coloured rings (placed nearest to front lens) to indicate immersion medium:
   Black – oil
   White – water
   Orange – glycerol
   Red – multi-media

[a]An excellent illustration of these markings is provided in the booklet by Kapitza (1994)

- to introduce graticules and pointers into a conjugate plane so that they may be seen in sharp focus at the same time as the image.

In theory, a single lens could serve these functions but in practice such a lens would need to be very large. This problem is overcome in the older types of eyepiece by adding a second lens – the field lens – to the eyepiece. The lens closest to the observer's eye is often called the 'eye' lens whilst that at the lower end of the eyepiece is the 'field' lens. A further problem which would occur if the eyepiece had only a single lens would be that the point of crossing of the emergent rays (the exit pupil or 'Ramsden disc') would be large. As the diameter of the pupil of the observer's eye, which should be located at the exit pupil of the eyepiece, has a maximum diameter of about 3 mm it follows that if the exit pupil of the eyepiece is larger than this there could be loss of some of the microscope field of view. This would be especially noticeable if the pupil of the observer's eye were not exactly at the exit pupil of the eyepiece.

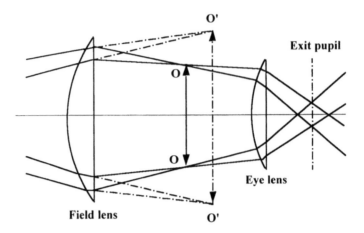

Figure 5.15. Effect of the field lens in the eyepiece. The field lens has the effect of reducing in size the real image in the primary image plane. This plane is brought nearer to the objective so that instead of a primary image of size O'–O' the actual image is O–O. This allows the observation of a larger field of view. The exit pupil or Ramsden disc is the point at which the rays cross above the eye lens of the eyepiece.

By adding a field lens the real image in the primary image plane is reduced in size, as shown in *Figure 5.15* where the image is effectively reduced in size from O'–O' to O–O and the primary image plane is moved slightly nearer the objective, so increasing the field of view.

There are two major types of eyepiece, known as internal or external diaphragm types, but other various special eyepieces are available. The internal diaphragm eyepiece (*Figure 5.16a*) has a diaphragm between the two plano-convex lenses which both have their curved surfaces facing the objective and are separated by a distance equivalent to twice the focal length of the upper or eye lens. These were originally called 'Huygenian' eyepieces. The lower, field lens has a focal length of approximately three times that of the eye lens. When uncorrected lenses are used together in this way as an eyepiece their spherical and chromatic aberrations tend to cancel out. The focal plane, where the primary image formed by the objective has to be placed, is inside the eyepiece and a fixed, field-limiting diaphragm is placed at this point. Such a simple eyepiece gives a satisfactory image with achromats but does not work well with semi-apochromatic or apochromatic objectives in microscopes using objectives corrected for a finite tube length. The chromatic difference of magnification of apochromats will not be corrected by the two uncorrected lenses of a simple internal diaphragm eyepiece.

The external diaphragm eyepiece (*Figure 5.16b*) also has plano-convex eye and field lenses but the field lens is mounted with its curved surface facing towards that of the eye lens. Typically in such an eyepiece (formerly often called a 'Ramsden' eyepiece) the field lens would have a focal length equal to that the eye lens and they would be separated by a

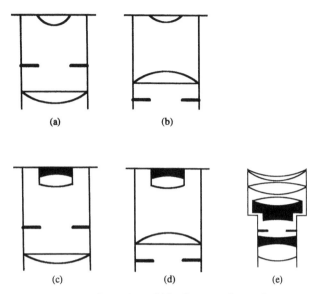

Figure 5.16. Different types of eyepiece. This diagram shows the construction of five different types of eyepiece: (a) internal diaphragm (Huygenian); (b) external diaphragm (Ramsden); (c and d) achromatized Huygenian and Ramsden (sometimes called a Kellner eyepiece); (e) modern 'Periplan' eyepiece.

distance equal to their focal length. The focal plane of the eyepiece lies outside the eyepiece, below the field lens, so that this feature makes this type of eyepiece useful for mounting graticules. They are more accessible and their image is clearer than if they are mounted inside, between the lenses of the eyepiece, as has to be done with a Huygenian eyepiece.

A version of the external diaphragm eyepiece, devised in the mid-nineteenth century by the optician Carl Kellner, used an achromatized doublet of crown and flint glass for the eye lens (*Figure 5.16d*). This achromatized Ramsden or Kellner eyepiece had both a higher eyepoint than the typical Ramsden eyepiece and also a larger field of view. Unfortunately, the field of view suffered from severe curvature which made these eyepieces unsuitable for photomicrography.

The problem of correcting the residual chromatic difference of magnification in the primary image of semi-apochromatic and apochromatic objectives (and high-magnification achromatic objectives as well) was solved by the use of specially computed eyepieces which introduced an equal and opposite error. These were called 'compensating' eyepieces. Because the errors deliberately introduced into the eyepiece must cancel those of the objective, it was essential that these two components should be made by the same maker. Failure to use the correct eyepiece with an older apochromatic objective designed for use at a finite tube length results in contrasty objects at the periphery of the field of view appearing with a red fringe on their outer diameters and a blue or violet fringe on the inner sides. Such compensating eyepieces originally gave adequate colour corrections but the flat portion of their field of view was very limited. This problem was solved in the 'Periplan' eyepiece, a variant on the Kellner eyepiece, which corrected the residual lateral chromatic aberration and, at the same time, gave a greatly increased area of usable flat field. *Figure 5.16e* illustrates the complex construction of a Periplan eyepiece with, in this case, six elements. With the introduction of 'plan' objectives in which the primary image had much less curvature of field that that of the older objectives and with the introduction of much wider body tubes to microscopes, the eyepieces had to cope with much larger primary images and this led to the development of specific 'wide-field' eyepieces by most makers, which allowed an area of up to 40% more of the specimen to be studied at any given time with a given objective.

With the use of infinity-corrected objectives, the tube lens usually assumes one or more of the former functions of the eyepiece, especially that of completing the correction of the residual errors in the objective.

All the bundles of rays leaving the eyepiece intersect at the exit pupil (Ramsden disc) or eyepoint (*Figure 5.15*). As mentioned above, this is where the pupil of the observer's eye should be placed in order to see the whole of the field of view. If the eye is above the exit pupil, then the peripheral rays cannot enter the eye and so the diameter of the observed field is reduced. As the magnification of the eyepiece is increased, so the exit pupil becomes located closer to the upper surface of the eye lens, which makes the eyepiece difficult to use, especially for spectacle wearers. This problem is avoided by the use of special 'high eyepoint' eyepieces which have been computed so that the exit pupil is located almost 20 mm above the top surface of the eye lens. The eye lens has to be of much larger diameter and extra components are needed in the construction. Spectacle wearers should use such eyepieces as they allow the users to retain their glasses (and hence obtain correction for both refractive and astigmatic errors) whilst using the microscope.

Eyepieces intended for measurement (often called 'micrometer' eyepieces) or for specific counting or sampling purposes, carry a graticule etched on to a circular glass disc in the plane of the field-limiting diaphragm of the eyepiece. In such eyepieces the eye lens has some form of focusing mechanism, either a helical thread or a sliding mount, to allow clear imaging of the graticule rulings. For more accurate measurements a 'filar micrometer' eyepiece is used; here a graticule with a rather coarse scale is mounted in the primary image plane together with a very thin wire. This wire is mounted so that it may be traversed across the field of view by a screw thread which carries an external calibrated drum. The wire is moved from one fixed graticule mark to the next by one complete turn of the drum, which (since it is divided into 100 divisions) allows the estimation of the wire movement within the major graticule divisions to be made with accuracy. Such a filar micrometer (or indeed any graticule eyepiece) must be calibrated for each objective with which it is to be used against an accurate ruling used on the stage as an object. This stage micrometer usually has a scale 1 mm long divided into tenths and hundredths of a millimetre. Full details of measurement with the microscope are to be found in Bradbury (1991).

To indicate specific objects in the microscope field a 'pointer' eyepiece may be used. This has a movable pointer mounted so that it appears in silhouette in the image plane. An improved version of this is the 'demonstration' eyepiece which has a beam-splitter mounted above the field lens and pointer. An offset side-arm with a second eye lens allows another observer to see the image with the pointer's silhouette at the same time. Even more elaborate systems are available in which the whole of the binocular head is duplicated and in which an arrow is projected optically into the image plane. For the study of a microscope image by more than two persons at once the former practice was to project a large real image on to a ground glass projection head. This suffered from the disadvantages that the image appeared degraded due to the grain of the ground glass and that the brightness of the image fell off markedly as the obliquity of the viewing angle increased. Such devices are now little used due to the ready availability of high-quality closed-circuit television systems with small CCD colour cameras.

Eyepieces were formerly available in a very wide range of magnifications although the present tendency is to have a more limited range available. In general the lowest power which allows adequate observation of the detail in the image is desirable, since the lower magnifications give a larger field of view. This diameter in millimetres is expressed as the field-of-view number or field number (FN). An 8× eyepiece might have a focal length of about 30 mm with a FN of 16 or 18 but special eyepieces might give a FN of over 25. If we know the field of view number, it is easy to calculate the real diameter of the object field from the formula:

$$\frac{FN}{M_{obj}q}$$

where FN is the field of view number in millimetres, M_{obj} is the objective magnification and q is any tube lens magnification factor in use. Hence with a FN of 18, an objective of 25× and a tube lens factor of 1.25 (due to a binocular head, say), then the actual area of the object field observed at any one time would be just over 0.57 mm.

Eyepiece magnifications of up to 25× are available but this would nowadays be regarded as excessive. The same total magnification (say 250×) could be obtained with a 10× objective with a NA of *c.*0.25 and a

25× eyepiece, or with a 25× objective of NA 0.65 and a 10× eyepiece. The image in the first case would be completely lacking in detail and appear hazy, whilst that from the objective with the higher magnification and higher numerical aperture would show the detail clearly.

Remember it is the numerical aperture of the system which provides the resolution. There is, however, a necessary minimum magnification needed to ensure that the detail resolved in the image may be seen by the eye. For this to occur the detail must usually be separated by an angle of at least 1 minute of arc; for this the minimum total magnification required is often arbitrarily quoted as (500 × NA). Further enlargement by the eyepiece will increase the angular separation of the detail in the image at the eye and make it easier to study. When the angular separation at the eye of two points in an image reaches about 2 minutes of arc (corresponding to a total magnification of about 1000 × NA), no further magnification will give any more information, assuming good acuity of the observer's vision. For older people a total magnification of about 1500 × NA is better. Increase beyond these arbitrary limits serves to make the image appear fuzzy and less clear. When this happens the limit of useful magnification has been exceeded and the image is said to suffer from 'empty magnification'. The value of 1000 × NA is again only arbitrary and it may be exceeded on occasions. For example, when counting micro-organisms, comfortable working may require that the magnification is greater than this arbitrary figure even though the absolute sharpness of the image may suffer.

References

Bradbury, S. (1991) *Basic Measurement Techniques for Light Microscopy.* Royal Microscopical Society Handbook No. 23. Oxford Scientific Publications, Oxford.

Bradbury, S. and Evennett, P.J. (1996) *Contrast Techniques in Light Microscopy.* Royal Microscopical Society Handbook No. 34. BIOS Scientific Publishers, Oxford.

Kapitza, H.G. (1994) *Microscopy from the Very Beginning.* Carl Zeiss, Oberkochen.

Pluta, M. (1988) *Advanced Light Microscopy, Volume 1: Principles and Basic Properties.* Elsevier, Amsterdam.

6 Illumination of the specimen

6.1 Light sources

In many old books on microscopy it was stated that a white cloud provided the ideal source of illumination; today an electric lamp would be used. If no integral illumination system is built into the microscope, the light source is often a mains-voltage opal electric bulb mounted in a simple housing and used with the microscope mirror. Lamps of this type are not entirely satisfactory for high power use because they are large, produce considerable heat and have a low luminance. If an external lamp is used, its housing need not have a collector lens but the hole through which the light emerges should be fitted with an iris diaphragm or a set of metal diaphragms with holes of different sizes. These will serve as the field diaphragm and control the size of the illuminated field when the microscope is used in the source-focused mode (see Section 6.4). If a focusable collector lens is fitted to an external lamp as described above, then Köhler illumination can be used (again, see Section 6.4). Some microscopes in the lower price ranges are fitted with low-voltage, low-wattage bulbs which have their filaments constructed with flattened, closely wound coils mounted side by side and facing the end of the glass envelope of the bulb. These give an extended source of light. Microscopes using this type of bulb may have a simple collector lens, often with a ground-glass undersurface to give diffusion of the light. Although the light bulb is built into the microscope base, these lamps are pre-aligned and the microscope is used in the source-focused mode. More expensive microscopes may have this same type of low-voltage bulb mounted in a pre-centred holder which may be moved with respect to the lamp collector lens for focusing the image of the filament in the front focal plane of the condenser (i.e. to give Köhler illumination).

Tungsten filament bulbs are powered from the mains via a transformer which is usually fitted with a voltage control. They emit a continuous

83

spectrum of light extending from about 300 to 1500 nm, with most of the energy output in the longer-wavelength region. This means that the light has a relatively low 'colour temperature' (i.e. it is somewhat reddish). As the lamp voltage is increased the luminance and the colour temperature of the lamp both increase markedly. The colour temperature of a light source is important in colour photomicrography (Bracegirdle and Bradbury, 1995) and is measured in kelvin (K).

Conventional bulbs suffer from the disadvantage that their light output declines as they age, as tungsten evaporates from the filament and condenses on the inner surface of the glass bulb, so blackening it.

Most microscopes are now fitted with tungsten–halogen bulbs (sometimes called quartz–iodine or QI bulbs) which have greater brightness and a higher colour temperature, when operated at their designed voltage, than conventional tungsten filament bulbs. Such tungsten–halogen bulbs (*Figure 6.1*) are very small and may be rated to give 100 W output at 12 V. They possess a single closely packed flat filament array which is mounted in a silica-glass envelope (often called 'fused quartz'); since they operate at a very high filament temperature (near to the melting point of tungsten) glass is not suitable for the bulbs as it would soften. The high filament temperature of the tungsten–halogen bulbs gives a very high luminance. One other characteristic of this type of bulb is that there is no blackening of the envelope as the bulbs age. The light output of such a bulb is remarkably constant throughout its life, although there

Figure 6.1. A selection of bulbs. From left to right: tungsten filament 6 V/20 W, accurately centred in special bayonet cap for the Wild M.20 microscope; tungsten filament 12 V/15 W, with small bayonet cap – the arrangement of the filament allows the bulb to be used side-on or end-on; tungsten filament 6 V/15 W, with bayonet cap – the filament is for use end-on, being a flattened rectangle facing forwards; tungsten filament 6 V/30 W – this bulb has the small Edison screw (SES) cap, is for use end-on, and is showing every sign of being about to burn out (pip on glass and blackening); tungsten filament 8 V/20 W – special design for use in Russian microscopes, with small bayonet cap, and intended for use end-on; tungsten–halogen 6 V/10 W, typical of the very small envelopes in which this output is available (compare with similar wattage tungsten filament bulbs); tungsten–halogen 12 V/50 W, with small envelope and very compact filament; tungsten–halogen 12 V/100 W, only slightly larger than the 50 W version and is standard in many units nowadays – this example has an especially large filament area and is for use in Olympus instruments; tungsten–halogen 12 V/100 W in dichroic reflector, much used in lamphouses for fibre-optic sources – the reflector allows most of the infrared radiation which causes heating to pass through, while reflecting back almost all of the shorter wavelengths.

may be some decline in the colour temperature for any given voltage as the bulb ages. Tungsten–halogen bulbs emit a continuous spectrum and when operated at their rated voltage have a colour temperature of about 3200–3250 K. Since tungsten–halogen bulbs run very hot, they need a heat filter in the beam path and they are mounted in a well-ventilated lamphouse attached to the stand.

For special purposes the light source may be a gas discharge tube filled with mercury or xenon vapour under pressure. All such lamps are of extremely high luminance. The mercury arc has a discontinuous spectrum with the emitted energy concentrated into peaks; those of wavelengths 365, 436 and 546 nm are particularly important. All three of these may provide the radiation for exciting fluorescence in a specimen. If a mercury arc is fitted with a narrow-band interference filter, the 546 nm band can be isolated to provide monochromatic green light. As mercury discharge lamps contain the vapour at a high pressure they have to be mounted in a substantial lamphouse which will also allow adequate cooling of the tube. They also need control units to provide the high voltage needed to strike the arc and to stabilize the current when the lamp is running. More rarely, xenon lamps are used. With this gas the light has a continuous spectrum and a very high colour temperature (5300–6000 K), approaching that of daylight.

6.2 The condenser

In transmitted-light microscopy the condenser:

- concentrates light on to the object with uniform intensity over the whole field;
- controls the aperture of the illuminating cone of light and matches it to the numerical aperture of the objective;
- can provide special types of illumination for phase contrast, darkground and other techniques.

In axial epi-illumination as commonly used today, the objective acts as its own condenser (see Section 6.6).

Condensers are often classified according to their purpose (e.g. whether they are for use with darkground or phase contrast). Alternatively, they may be considered in terms of the degree of their optical corrections. They would thus be uncorrected, or achromatic, or achromatic/aplanatic. Finally they may be grouped according to their maximum numerical aperture into low-power condensers (NA up to 0.25), medium-power dry condensers (working up to 0.95 NA) and high-power oil-immersion condensers (NA of up to 1.4).

The design of condensers must be a compromise between the need to provide a cone of light suitable for objectives with different numerical apertures (from very low to the maximum of 1.4) and the production of a focused image of the illuminated field diaphragm which is large enough to fill the whole of the visible objective field. With a low-power lens, say of magnification 2.5×, this may be a circle of about 7 mm diameter whilst with the 100× oil-immersion lens the diameter of the illuminated field will be about 0.2 mm. The maximum illuminated area produced by a condenser is a function of its focal length: the larger the area to be covered, the longer the focal length must be and the smaller the maximum aperture of the cone of light which it can produce. Low-power objectives need to have a large area illuminated with a cone of low aperture. Ideally this is provided by a specially designed condenser of long focus. High-magnification objectives, on the other hand, require a small area to be illuminated with a cone of light of large aperture for which a short-focal-length condenser is needed. Many of the condensers used with the cheaper modern microscopes are designed for use with objectives of the middle range of powers and will neither illuminate a sufficiently large field for the lowest magnifications nor produce an adequate cone of rays for the very highest aperture oil-immersion objectives.

Ideally separate condensers should be used for the very low- and very high-aperture objectives. In practice the problem is often solved by providing the condenser with an accessory lens which may be added below the main elements to allow it to be used with very low-power objectives; alternatively, the top lens may be swung out of the optical path. Both of these arrangements effectively change the focal length of the condenser. When the condenser is used without its top lens, the aperture diaphragm (which is located in the front focal plane of the complete condenser) will no longer be in this plane, and the aperture diaphragm may not function as such and the illuminated field diaphragm may not act as intended, which is to limit the area of the illuminated field and control glare. This may be of little significance, since with the much lower NA of a condenser used without its top lens, the light rays pass through the object plane at angles much closer to the axis and the percentage of reflections which may cause glare in the image is much reduced. With the top lens removed the conditions for true Köhler illumination may not be fulfilled; the lamp iris diaphragm must now be imaged into the back focal plane of the objective by the condenser and objective acting together. The lamp iris thus acts as the illuminating aperture diaphragm for the system, with the object plane illuminated by a series of cones of light of equal angles of convergence. In such cases the iris diaphragm in the condenser must be fully opened in order to prevent fall-off in the intensity of the light towards the periphery of the field of view. For this reason some condensers designed for use with very low-power objectives are not fitted with an aperture iris.

If a numerical aperture of greater than 0.95 is needed from a high-quality condenser, the upper lens of the condenser must be linked to the underside of the microscope slide by a film of immersion oil. This ensures that the extremely oblique rays of light which exceed the critical angle can escape from the condenser and are not reflected from the underside of the slide. Oiling the condenser is troublesome (as well as messy) and is seldom needed in routine microscopy. Work at the limit of resolution with apochromatic objectives is an exception. Note that when using high-power darkground condensers oiling to the slide is essential.

As the condenser has to focus an image of the illuminated field diaphragm into the plane of the specimen, the substage mount for the condenser must have some means of focusing as well as for centration to the optical axis of the microscope. Almost all condensers have an iris diaphragm at or close to their front focal plane. This diaphragm is the illuminating aperture diaphragm (the 'aperture diaphragm' for short) and it limits the NA of the cone of light delivered by the condenser to one suitable for the objective in use. On older instruments it was also usual to provide one or more swing-out holders to the underside of the substage to carry colour filters for contrast enhancement or colour temperature correction in colour photography.

6.3 Condenser types

The condenser most commonly used in cheaper microscopes is the two-lens illuminator, devised by Ernst Abbe, which is shown in diagrammatic section in *Figure 6.2a*. This condenser is uncorrected for chromatic and spherical aberrations and so it is not suitable for use with the highest-quality objectives. The Abbe condenser produces a cone of rays which at

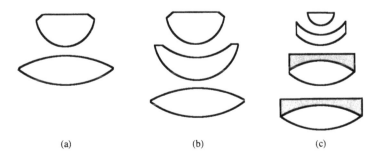

(a) (b) (c)

Figure 6.2. Lens arrangement in three types of condenser: (a) a simple uncorrected Abbe two-lens illuminator; (b) an aplanatic Abbe condenser has one extra meniscus lens; (c) an achromatic-aplanatic condenser has at least two corrected doublet elements.

full aperture does not come to a sharp focus (because of the spherical aberration) but shows an extensive zone of confusion (just as shown in *Figure 1.13* for an objective). This means that the image of the illuminated field diaphragm formed by an Abbe condenser is not very sharp and is surrounded by colour fringes (because of the lack of colour correction in the condenser). A typical Abbe two-lens illuminator, used 'dry', would have a maximum aperture of about 0.5. The chief advantages of the Abbe condenser are its cheapness and long working distance, and the ability to illuminate a large area of the object.

Better results are obtainable with more highly corrected condensers. The Abbe three-lens or aplanatic condenser (*Figure 6.2b*) is better corrected for spherical aberration than the standard Abbe but is still uncorrected for colour. It has a 'dry' NA of about 0.65 and may be used oiled to the underside of the slide, when its aperture becomes about 1.2. Despite its lack of colour correction this type of condenser is much used in routine microscopy. The most expensive achromatic/aplanatic condensers (*Figure 6.2c*) are also chromatically corrected for at least two wavelengths. When oiled to the slide such condensers can give an aplanatic cone up to a NA of about 1.4. Such a condenser would be needed to obtain maximal resolution from a high-aperture apochromatic oil-immersion objective.

6.4   Source-focused and Köhler illumination

Controlled illumination of the specimen is essential. Control of the illumination ensures comfortable viewing, even lighting of the specimen and limits the area of the specimen which is illuminated. It is good practice to illuminate only that part of the specimen which is visible, as this minimizes glare from scattered light which reduces the contrast of the image.

Over the years, two principal arrangements of lamp and condenser have evolved to satisfy these needs. The older method, termed 'source-focused' illumination (or sometimes 'critical' or 'Nelsonian' illumination, after its chief supporter), was used in the last century with the flame of an oil lamp as the source. Source-focused illumination is still used extensively in the cheaper student microscopes sold today and by many amateurs who use older stands with an external lamp. A homogeneous light source such as that of an opal enlarger bulb is imaged directly by the condenser into the plane of the microscope specimen (*Figure 6.3*). The area of the specimen illuminated is controlled by an illuminated field diaphragm (or a series of fixed apertures of varying sizes) placed just in front of the lamp. According to the series of conjugate planes described in Chapter 5 it follows that there will be a focused image of the lamp surface, the field

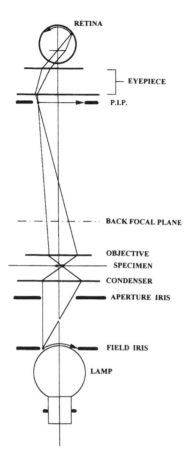

Figure 6.3. Arrangement of source-focused illumination. The bulb in this type of illumination (sometimes also called 'Nelsonian' or 'critical' illumination) should provide a completely homogeneous source of light. 'Flashed opal' bulbs, as used in photographic enlargers, are often used. There is usually no collector lens in front of the bulb.

diaphragm and the specimen in both the primary image plane and on the retina of the eye. For this reason the light source must be large, structureless and have an even intensity over its surface. Source-focused illumination is adequate for much routine visual microscopy when an external light source may have to be used. It is simple and easy to set up.

Most modern microscopes with integral illumination use a low-voltage filament lamp or a tungsten–halogen lamp (see above). Both have a tungsten filament which has a marked structure and so such lamps cannot be used in the source-focused mode. Instead, a technique devised originally for photomicrography by August Köhler is used. A collector lens is placed in front of the lamp with the filament located near its focus. The illuminated field diaphragm (commonly called the 'field iris') is mounted in front of the collector lens so that the microscope condenser will focus the image of the illuminated field diaphragm into the specimen plane (*Figure 5.4*). The lamp collector lens projects an image of the filament into the front focal plane of the condenser where the illuminating aperture diaphragm (the 'aperture iris' or just the 'iris') is located (*Figure 7.1*). Thus an image of the lamp filament will be found at each conjugate plane in the 'aperture' series (i.e. in the back focal plane of the objective and the exit pupil of the eyepiece; *Figure 5.4*). The filament image will appear nowhere else. The lamp collector lens appears as a homogeneous secondary source which, together with the image of the illuminated field diaphragm, will be imaged into the specimen plane. This will thus appear to be illuminated evenly by a source of controllable size. Images of the field diaphragm and the specimen will be found in each succeeding conjugate plane of the 'field' series, namely the primary image plane, and on the retina of the eye. The ray paths for each of the two sets of conjugate planes are shown in *Figure 5.4*; note that the image-forming rays have been arbitrarily separated from the rays in the illuminating path and shown on the left and right of the diagram respectively. Details of the adjustment of a microscope to achieve Köhler illumination will be found in Chapter 7.

For Köhler illumination the lamphousing must be fitted with a collector lens, often including an aspherical element. Such collector lens systems must be capable of projecting an enlarged image of the lamp filament of such a size that the whole of the maximum working aperture of the condenser is filled with light. In a Köhler system a diaphragm is fitted close to the lamp collector lens to serve as the illuminated field diaphragm. This acts firstly to alter the diameter of the illuminated area in the object plane. Light from parts of the specimen outside the visual field may be scattered into the objective and light from the periphery of the primary image falling on the inside of the microscope tube will be scattered into the eyepiece (so causing glare and image degradation) unless the image of the illuminated field diaphragm is restricted in size to the area under observation. In addition, secondly this acts as a means, when partly closed, of allowing the focus of the microscope condenser to be set and the centration of the condenser to be checked. When the condenser is focused and centred a sharp image of the field diaphragm will be seen at the same time as a sharp image of the specimen and concentric with the boundary of the field of view (see *Figure 6.4*).

Some microscopes have a mount for holding colour filters near the illuminated field diaphragm. This position is not ideal, since any surface blemishes on the filter will be imaged into the specimen plane and hence will be clearly visible. Colour filters are best mounted well away from any plane conjugate with the image; an alternative position such as just under the aperture diaphragm of the condenser is more satisfactory.

As well as controlling the illuminated area of the specimen it is even more important to control the angle of the cone of light used; the size of angle required is governed by the numerical aperture of the objective.

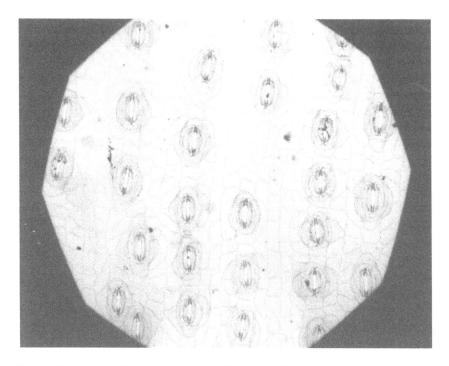

Figure 6.4. Imaging the illuminated field diaphragm. This photograph was made of the whole field produced by a 10× plan-apochromatic objective, with the illuminated field diaphragm closed somewhat. The sharp image of the edge of this diaphragm demonstrates two things. One is that the substage condenser is in focus, the other is that the substage condenser is highly corrected. If either condition is not met, the image of the diaphragm will be unsharp.

Although most texts recommend a cone of light which occupies 70–90% of the aperture, with the very best of today's objectives the corrections are so good that a 95% cone may be used. If a much smaller cone is used, then the maximum resolution attainable is markedly reduced, the image appears dark, and all fine detail is surrounded by prominent diffraction lines (see *Figure 6.5*). This effect is sometimes termed a 'rotten image'; such an image is often preferred by beginners because it has an increased contrast which gives the mistaken impression that more detail is actually visible, rather than the reverse! As mentioned, use of too large a cone of light causes glare and consequent loss of contrast.

From the above points it is clear that the illumination must be arranged so that:

- the angle of the cone of light from the condenser entering the objective is controllable by means of an aperture diaphragm;
- the direction of the cone of light is controllable by movement of the diaphragm;
- the area of the illuminated field is controllable by an illuminated field diaphragm.

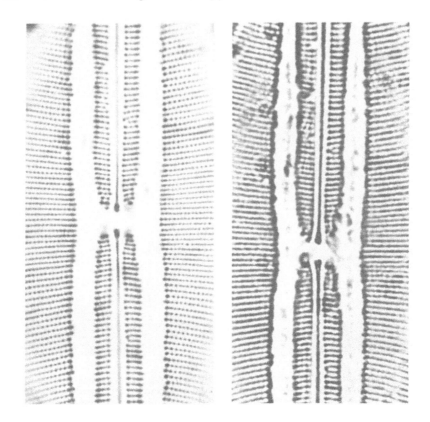

Figure 6.5. Effect of diffraction on the image. Two photographs of the diatom *Navicula lyra* both taken with a 40× objective of NA 0.75. On the left, with the field and aperture diaphragms correctly adjusted, fine detail is resolved. The right-hand picture shows the effect of closing the aperture diaphragm in the condenser as far as possible. The contrast is greatly increased but the resolution is destroyed by diffraction effects. This is an example of a 'rotten' image.

The amount of the back focal plane of the objective which is illuminated by the cone of light from the condenser (controlled by the angle of the illuminating cone) is observed by removing the eyepiece and looking down the tube of the microscope. If a Bertrand lens is fitted in the microscope tube, then this, when inserted, acts together with the eyepiece to form a telescope which performs the same function.

6.5 Transmitted illumination

If the preparations to be examined are relatively thin it is possible for light to pass through them without too much being absorbed. They may thus be examined by transmitted light. Transmitted light is used for many biological and histological specimens and for thin sections of rocks and minerals. In normal practice the light provided by the condenser forms an axial cone of rays which occupy most of the aperture of the condenser and objective. This is called 'bright-field' illumination and works well with stained biological or material specimens where the contrast is high, although if the contrast is low (as in *Figure 6.6a*) it is not very satisfactory. For some special purposes (e.g. to obtain resolution of very fine detail; see Chapter 5), or to obtain extra contrast, a narrow cone of rays which are deliberately made off-axis may be used. This type of illumination is called 'oblique' (*Figure 6.6b*) and it produces images which are not easy to interpret because the lighting all appears to come from one side. Instead of the obliquity being only in one azimuth it is possible to use oblique light in two azimuths orientated at right angles to one another. This form of lighting was much used by the students of diatom morphology 100 years ago.

If the lighting is oblique in all azimuths and its obliquity is such that no direct beam can enter the aperture of the objective, then we have 'darkground' illumination (*Figure 6.6c*). Darkground illumination is a valuable technique for increasing the contrast of unstained specimens. In this technique direct light is not allowed to take part in the formation of the image; this is done either by blocking the zero-order light in the back focal plane of the objective or, more usually, by illuminating the specimen with a hollow cone of light of such obliquity that all the light falls outside the acceptance angle of the objective (*Figure 6.7*). This is simple with low-power objectives. A central opaque stop (a 'patch stop' – see *Figure 6.8*) of the correct size is fitted in the front focal plane of the condenser; this ensures that all the light falling on the object is too oblique to enter the objective. Light diffracted by the specimen, however, emerges in a cone over 180 degrees (this is shown in *Figure 6.8*); some of this will enter the objective and form an image in which features scattering light will appear bright. Patch stops are most effective with objectives of NA less than about 0.65. For objectives of higher aperture, reflecting condensers are used instead, oiled to the slide. Such condensers may have a single reflecting surface formed as part of a paraboloid or they may have two reflecting surfaces, either both spherical or with one formed as part of a cardioid surface. Even with such condensers it may not be possible to obtain an adequate darkground when a high-aperture oil-immersion objective is in use. This problem may be resolved by restricting the numerical aperture of the objective by a small iris diaphragm built into its back focal plane. Some resolution will be sacrificed but this is usually less important than the gain in contrast achieved by satisfactory darkground. Darkground reflecting condensers need care in their centration and focusing so that the specimen is located exactly at the focal point of the cone of rays.

Phase contrast is a technique intended to enhance the contrast of specimens which have little absorption of light, for example living biological

(a)

(b)

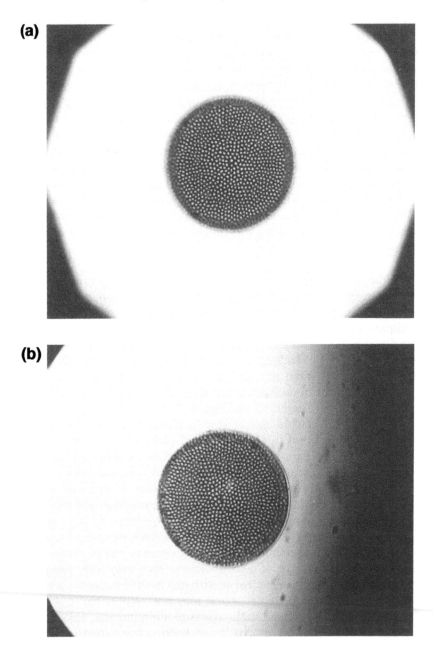

Figure 6.6. Three modes of illumination. (a) Bright-field illumination. A diatom mounted in a medium of refractive index close to that of the frustule is imaged using a 40× objective and corrected condenser. Even with the illuminating aperture diaphragm much reduced, the contrast is low. (b) Oblique illumination. The same diatom at the same focus with the same objective is now imaged with a sector stop introduced below the same condenser as in (a). This stop is a metal disc with a segment cut out, providing illumination from one azimuth only (above, as reproduced here). Although the contrast

(c)

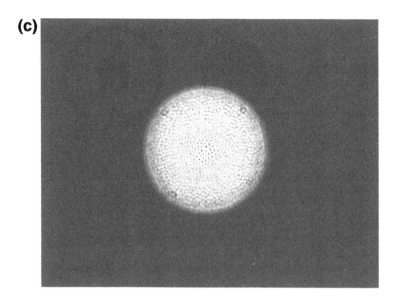

is much higher, the resulting image must be interpreted with caution, for such treatment gives a 'pseudo-relief' effect which is really an optical artifact; the method does, however, reveal details which can later be investigated in a more conventional manner (e.g. by Nomarski differential interference contrast). (c) Darkground illumination. The same diatom (at the same focus with the same objective) is imaged with a darkground condenser. Scattering features appear bright, and contrast is reversed and much higher.

material. This technique requires modifications to both objectives and condensers and is dealt with in Bradbury and Evennett (1996). The condensers intended for use with phase-contrast objectives are fitted with a rotatable disc in the front focal plane of the condenser. This disc carries

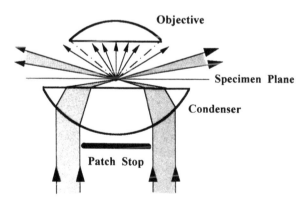

Figure 6.7. Principle of darkground illumination. A central patch stop in the condenser occludes all the light which would otherwise fall within the acceptance angle of the objective. If no specimen were present the field of view would be dark. The hollow cone of light from the condenser (shown stippled) has the object at its focus. Diffracted light (small black arrows) from the specimen enters the objective and forms the image in which light-scattering features appear bright on a dark background.

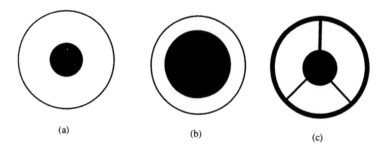

(a)   (b)   (c)

Figure 6.8. Darkground patch stops to be placed in the condenser filter holder. (a) and (b) are constructed on a circle of clear glass or plastic by adding a central circle made of black Indian ink or tape. The smaller circle (a) would be for a small-aperture, low-power objective whilst the larger (b) is for an objective of higher NA. The sizes of the occluding circles are adjusted empirically to suit each individual objective. Diagram (c) represents a commercial patch stop made of metal with the central stop supported by three arms.

not only the aperture diaphragm but also a series of annuli of different sizes which are matched to the phase-plates in the objectives. Because they are in a plane conjugate with the back focal plane of the objective, these annuli are imaged sharply on to the ring in the phase-plates. This ensures that the hollow cone of the direct light used in phase-contrast microscopy all passes through the correct region of the phase-plate. Each annulus must be centrable to the phase-plate in its own objective, independently of the centration of the condenser as a whole.

6.6   Epi-illumination

If the specimens are so thick that light cannot pass through them, for example polished blocks of metal, or if they are very intensely coloured, then they must be examined with light incident on them from above. This is often called epi-illumination. In the simplest form of operation with low powers (as with a stereomicroscope) one or two spot lamps mounted above and at an angle to the microscope stage may be used to shine light directly at an angle on to the specimen. With higher powers of the compound microscope (say 10× upwards) glare is more troublesome than with transmitted light and careful adjustment of the system is needed for the best results. With microscopes designed for epi-illumination the light enters the tube at right angles to the optical axis and is reflected down through the objective (which then acts as its own condenser) on to the specimen by a semi-transparent reflector mounted above the lens elements of the objective at 45 degrees to the optical axis. In some old microscopes a simple adaptor was available for this purpose which contained a very thin plane piece of glass; occasionally this glass, which resembled a small circular coverslip, was partially silvered. Another variant used a

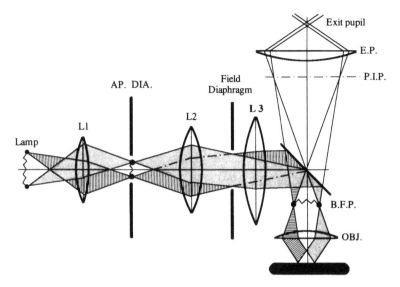

Figure 6.9. Epi-illumination system. A ray diagram of Köhler illumination with a reflected-light system. L 1 is a relay lens and L 2 is the lamp collector lens; AP. DIA. is the aperture diaphragm; B.F.P. is the back focal plane of the objective, OBJ.; P.I.P. is the primary image plane of the system; and the eyepiece is labelled E.P. Note that, in contrast with a transmitted-light system, the positions of the field and aperture diaphragms are reversed, the latter having been transposed to the preceding conjugate plane in the aperture set. The angles of the rays are much exaggerated for clarity; the crossing of four rays at a single point on the semi-silvered reflector is coincidental and is arranged thus for ease of drawing.

small prism which occupied part of the back focal plane; with this the incident light reached the specimen at a slightly oblique angle, and there was a reduction in the working numerical aperture of the objective. Modern systems use a beam-splitting cube and allow full control of Köhler illumination for the specimen. As described above for transmitted light, oblique illumination may also be used with epi-illumination both in bright-field and in darkground.

With modern reflected-light illuminators a relay lens (L 1 in *Figure 6.9*) is included in the optical illuminating train. This projects an image of the lamp filament into the plane of the illuminating aperture diaphragm. An image of both of these is projected into the back focal plane of the objective by lens L 2 (*Figure 6.9*), the equivalent of the lamp collector lens of a transmitted-light system. This allows control of the illuminating aperture by closure of the aperture diaphragm *without* affecting the aperture of the objective lens with respect to the light reflected back from the specimen. If the aperture diaphragm were to be physically placed in the back focal plane of the objective this independent adjustment would not be possible.

Many biological applications of microscopy now involve fluorescence and confocal microscopy, both of which are usually carried out with epi-illumination and in which the emitted imaging light is often of low intensity and in which maximum detail is often required to be resolved. This is also true for the microscopy of weakly reflecting materials such as coal. All of these need a maximum aperture of the objective lens. The illuminated field diaphragm is mounted so that the objective projects its image into the specimen plane. Control of the illuminated area of a low-reflectance specimen in order to reduce glare is even more important with epi-illumination than with transmitted light. Note that with a reflected-light Köhler system as described here the relative positions of the field and aperture diaphragms are reversed relative to the light source, the aperture diaphragm being nearer to the lamp (compare *Figures 6.3* and *6.9*). Because it is not desirable to position a diaphragm in the objective back focal plane (where it would also restrict the imaging aperture), the diaphragm is placed in the preceding conjugate plane.

References

Bracegirdle, B. and Bradbury, S. (1995) *Modern PhotoMICROgraphy.* Royal Microscopical Society Handbook No. 33. BIOS Scientific Publishers, Oxford.
Bradbury, S. and Evennett, P.J. (1996) *Contrast Techniques in Light Microscopy.* Royal Microscopical Society Handbook No. 34. BIOS Scientific Publishers, Oxford.

7 Practical use of the microscope

7.1 Getting a good image

It is unfortunate that however badly a microscope is used, it almost always gives an image, although the quality of this may leave a great deal to be desired. It is very easy to set up the instrument correctly so that it will give the best image of which it is capable. If certain simple rules are followed it takes only a very short time to obtain an image of high quality. The detailed study of any microscopical specimen may take a considerable amount of time so before starting it is a good idea to make sure you are comfortable and that your seat is of such a height that you can look into the eyepieces of the microscope without having either to extend your neck or to stoop. It is also a good idea to rest your forearms on the bench so that the microscope stage movements and the focus controls may be operated easily and comfortably.

There are so many differing models of microscope in use that it is not possible to give detailed instructions in one book: *the user should consult the instruction manual* for the particular microscope that is in use. The following instructions give a generalized outline and provide a basis for setting up a typical microscope for use with transmitted light and Köhler illumination.

(i)   Place a contrasty stained specimen on the stage, rotate the nosepiece until the $10\times$ or $16\times$ objective is in position and switch on the light. Open all diaphragms and raise the condenser to the top of its travel.

(ii)   If the microscope has a binocular head check the interocular separation of the eyepiece tubes and reset if necessary.

(iii)   Focus the specimen with the coarse focus adjustment. This is best done by first bringing the front of the objective to about 5 mm from the surface of the slide, watching from the side at stage level. Then, looking through the eyepieces, increase the distance between the specimen and the objective until the image appears sharp. Make the final adjustment with the fine focus of the microscope.

(iv)   If you are using a binocular microscope, it may be necessary to adjust the individual eyepieces for variations in your own eyes. At least one eyepiece will have an adjustment collar. Close the eye which is using the adjustable eyepiece and, using the other eye, focus the microscope sharply on some detail in the image. Now use the other eye and focus the adjustable *eyepiece* by turning its collar until the same detail in the image appears as sharp as before.

(v)   Check that the illuminating aperture diaphragm (the condenser iris diaphragm) is wide open. Partially close the illuminated field diaphragm until its image (probably blurred and possibly off-centre) appears in the field of view.

(vi)   Make the image of the illuminated field diaphragm as sharp as possible by altering the focus of the *condenser*. If the condenser is an Abbe, it will be likely that this image will be fringed with colour. This is unavoidable; aplanatic and achromatic condensers, however, will give much sharper images of the field diaphragm.

(vii)   If necessary, centre the image of the illuminated field diaphragm. This is made easier if the diameter of the image of the field diaphragm is made almost coincident with that of the field of view. On most instruments the illuminating system, including the illuminated field iris, is aligned during manufacture and the centration of the image of the field iris is done with the condenser centring screws. In some cases the diaphragm may be adjusted with its own centring screws.

(viii)   Open the illuminated field diaphragm until the whole of the field of view is illuminated.

(ix)   Remove one of the eyepieces and look at the back focal plane of the objective. If a magnifier provided with phase-contrast kits (often called a 'phase telescope') is available this is a help in seeing the focal plane clearly. Close the illuminating aperture diaphragm in the condenser until its image just encroaches on to the illuminated area in the back focal plane.

(x)   Replace the eyepiece and adjust the brightness of the light to a comfortable level by using the brightness control or by inserting neutral density filters into the light path.

The microscope will now give the best image of which it is capable. If the objective in use is not well corrected then it may be necessary at stage (ix) above to close the illuminating aperture iris rather more to obtain a satisfactory image – with an old lens perhaps even reducing the working aperture of the objective by a third of its diameter. Only experience is a good guide in these circumstances.

7.2 Changing to higher-aperture (higher-magnification) objectives

The essential difference between the low-power objectives and those of higher power is that the latter have much higher numerical apertures and hence greater resolving power. This means that on changing from low to high power it is necessary to open the condenser illuminating aperture iris so as to obtain a sufficiently large angle of the cone of light to give adequate resolution of detail by the objective. It cannot be stressed too emphatically that each objective should be operated with a condenser cone large enough to fill 70–90% of the aperture of the objective lens. Failure to use a suitable condenser aperture (i.e. the cone of light is too small) will give a 'rotten' image, lacking in detail and with diffraction haloes around the object detail as illustrated in *Figure 6.5*. On returning to a lower-power objective the setting should be readjusted; too large an illuminating cone will often produce 'glare' which will seriously reduce the contrast of the image.

7.3 Using a very low-power objective

It is a good principle to do as much observation as possible at low magnification (see Chapter 1). This allows the observer to relate the microscopic morphology much more easily to the structures which have been seen with the naked eye or the hand lens. Only when all the possible information has been obtained from one power should the next higher be used. Students and others beginning to use the microscope often move on to the higher powers too rapidly and so make their task of interpreting the object structure far more difficult than it should be. When using a low-power objective (e.g. one with a magnification of 4×) the numerical aperture of the condenser must of course be adjusted to suit that of the objective; it is probable that the condenser will not illuminate the whole field of view, even when the field diaphragm is fully open. If this is the case then either the top lens of the condenser must be removed (sometimes it is designed to pivot out of the optical axis for this purpose) or the condenser must be changed for one of longer focal length. When special objectives designed to give a magnification of 1× or 1.5× are to be used, a matching condenser is essential.

7.4   Oil-immersion objectives

For studies of detail near the limit of resolution, oil-immersion objectives with numerical apertures greater than 1 are used. The method of use is as follows.

(i)   Locate the object of interest with a lower-power objective, place it in the centre of the field of view and focus.

(ii)   Place a single drop of immersion oil on the top of the coverslip and, assuming that you are using modern parfocal objectives, slowly swing the oil-immersion objective into place. Make sure that its front lens makes contact with the drop of oil.

(iii) Look through the microscope; the preparation should be roughly in focus. Use the fine focus to make the image sharp.

(iv)   If the condenser is being used 'dry', then check that its aperture iris diaphragm is fully open. In such a case the oil-immersion objective cannot be operated at its maximum illuminating aperture because the condenser can only deliver a cone of maximum NA of 1.0. Although rarely required, if you must get the maximum possible resolution in transmitted light from an oil-immersion objective of very high NA, then the top lens of the condenser also must be oiled to the underside of the slide (a special, high-aperture oil-immersion condenser is desirable). After focusing the condenser, its aperture is adjusted to match that of the objective.

For the best results, only that small area of the specimen which is actually under observation should be illuminated. As the magnification of the objectives is increased, then so the size of the illuminated field aperture must be reduced, otherwise the image will be degraded by glare. This is especially important when using incident (reflected) light. Another possible cause of a poor image is the presence of bubbles in the drop of oil. These may be seen if the eyepiece is removed and the back focal plane of the objective inspected. Bubbles may often be dislodged by moving the slide to and fro or by partially rotating the nosepiece. If these tricks are not successful in clearing the bubbles then the oil must be wiped off and replaced with a fresh drop.

7.5   Darkground

Darkground illumination is a very valuable technique for increasing contrast, the principle of which is illustrated in *Figure 6.7* and, provided simple precautions are taken, it is not difficult to achieve. Darkground is discussed here because it is so simple, gives such useful information about transparent specimens and its use is very often overlooked by the 'modern' microscopist. For low powers (up to say 20× initial objective magnification) an ordinary Abbe or achromatic condenser may be used, fitted with a central opaque 'patch stop' (see *Figures 6.7* and *6.8*). For higher powers a special reflecting darkground condenser is needed, which must be in oil-immersion contact with the underside of the slide. With such condensers particular care must be paid to their centration. Some darkground condensers have a small circle engraved on their upper surface to help with this centring. The slide is removed, the circle imaged with a low-power objective and the centring screws of the condenser adjusted until the image of the circle is concentric in the image field. The slide may then be replaced and the condenser oiled to the underside of the slide before the correct objective is focused on to the object. Finally the condenser focus is altered until the best darkground effect is achieved, with the object standing out brilliantly illuminated against a completely black background field. If a high-power darkground condenser lacks the centring circle it is usually possible to centre it by focusing a low-power objective on to the slide and bringing the condenser to its point of focus (after oiling it to the underside of the slide). The focal point will be seen as a bright spot of light which must be centred in the field of view by the adjusting screws. The high-power objective is then substituted and final adjustments to the condenser focus carried out. Note that if an Abbe or an ordinary achromatic condenser is used with a central patch stop for darkground illumination then it is *essential* that the aperture diaphragm is fully opened so that the peripheral rays (which illuminate the specimen at a very oblique angle) may pass. Reflecting darkground condensers are not fitted with an aperture diaphragm.

When using darkground it is important to ensure that all optical surfaces in the light path (e.g. slide, coverslip and the mountant itself) are clean and free from dirt and inclusions such as air bubbles. Any such foreign matter will scatter light very effectively and provide obtrusive out-of-focus patches of light which effectively destroy the image contrast.

7.6 Reflected light or epi-illumination

If opaque objects are to be examined the light is provided from a source above the microscope stage. With stereomicroscopes and at low powers of the standard instrument (up to objectives of a magnification of 10×, say) no special modifications of the microscope are needed. Low-magnification objectives have a long working distance and it is thus possible to use separate spotlights positioned at the side of the stage and focused upon the specimen. Such illumination will, of course, be off the optical axis of the microscope but in many cases this is an advantage as it provides more information about the surface relief of the object. If axial light is required and for higher powers a special objective and lamp unit must be used to provide Köhler illumination. Glare is often more troublesome than with transmitted light and careful adjustment of the illumination system is essential for good results. The light is reflected on to the specimen through the objective lens (which thus acts as its own condenser) by a semi-transparent reflector mounted above the back focal plane of the objective. In some simple attachments the reflector may be a very thin piece of plane glass or a partially silvered piece of glass. Occasionally a very small prism is used mounted to one side of the optical axis, but with this arrangement there is inevitably some obliquity of the light and a reduction in the maximum working aperture of the objective. Current equipment intended for reflected light would use a beam-splitting cube and a built-in illuminating optical train using Köhler's principles. With the system illustrated diagrammatically in *Figure 6.9*, the lens (L 1) transfers an image of the lamp filament into the plane of the illuminating aperture diaphragm (AP. DIA. in *Figure 6.9*). The lens L 2 (the equivalent of the lamp collector in transmitted light) projects any image of the lamp filament and the aperture diaphragm into the back focal plane of the objective. This allows control of the illuminating aperture by closing the aperture iris *without* affecting the imaging aperture of the objective lens with respect to the light refracted from the specimen. If the illuminating aperture diaphragm were to be placed in the back focal plane of the objective itself such an independent adjustment would not be possible. The illuminated field iris (field diaphragm) is placed close to lens L 2 so that this lens and the objective, acting together, project the diaphragm image into the specimen plane. Control of the illuminated area of specimens illuminated with incident light so as to reduce glare is even more important than when transmitted light is used.

7.7   Lamp adjustment

When Köhler illumination is to be used with a microscope and a separate external lamp, the lamphouse must have a collector lens. It should be carefully aligned with the microscope (preferably by mounting them together on a microscope board) and the image of the filament centred and focused by the lamp collector lens on to the plane of the aperture diaphragm of the microscope condenser. The aperture diaphragm of the microscope condenser is temporarily closed fully to allow this to be done; the appearance should be as shown in *Figure 7.1*.

In some microscopes fitted with built-in illumination there is provision for adjusting the position of the bulb with respect to the optical axis of

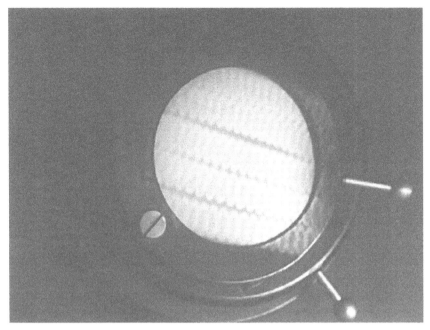

Figure 7.1. Filament image: the appearance of the 'in focus' image of the filament of an external lamp on the condenser diaphragm. Here the whole aperture of the condenser is occupied by the filament image.

the collector lenses. If no mirror is fitted behind the bulb and there is no ground-glass diffuser in the collector lens train, then it should be possible to obtain an image of the lamp filament by placing a piece of ground glass or white card just below the condenser illuminating aperture diaphragm. This plane may be observed without discomfort by using a small piece of mirror held at an angle of 45 degrees under the condenser. When the lamp filament is correctly centred its image should be as in *Figure 7.2a*.

If a concave mirror is mounted behind the bulb to give a double image of the filament (thereby effectively providing a much larger primary light source), then on first inspection the appearance might be as *Figure 7.2b*. Here two images of the filament are seen, one being formed via the mirror. The adjusting screws on the bulb holder should then be manipulated until the images appear in line and just touching (*Figure 7.2c* and *d*). If a conventional solid source filament bulb is fitted it may be possible to remove it together with its collector lens system and project an image of the filament on to a screen. Any necessary centration is then apparent and errors may be corrected as before by using the adjusting screws on the lamp housing.

If an external light source is used for Köhler illumination with a microscope fitted with a substage mirror, the image of the lamp filament must be focused with the lamp collector lens on to the under surface of the

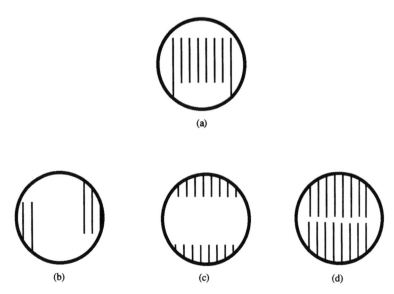

Figure 7.2. Diagrams of the appearance of the filament image of an internally mounted tungsten–halogen bulb. (a) The single image of an internally mounted tungsten–halogen bulb in the aperture diaphragm plane of the condenser. This often does not completely fill the aperture. (b) One possible appearance when the illumination system is out of alignment. Two images of the filament (one inverted and produced by the mirror behind the bulb) are not coincident. (c) The two images have been aligned but much of the aperture is unfilled with light. (d) The final correct adjustment, when the images have been aligned and brought close together.

condenser aperture diaphragm. This may easily be inspected by the use of a small accessory mirror held in the hand. When the lamp is correctly adjusted the appearance should be as in *Figures 7.1 or 7.2a, d.*

7.8 Setting up source-focused illumination

This method of illumination is often used when the microscope is fitted with a mirror and an external light source is used. The lamp should be mounted in a suitable screened housing which is fitted with some means of limiting the area in front of the bulb. This may be an iris diaphragm but a series of interchangeable stops of differing sizes, made from thin, blackened metal, are equally suitable. They should be mounted as close to the bulb as possible. The light bulb itself should be sufficiently large and homogeneous so as to illuminate the whole field of view of the lowest-power objective to be used. The so-called 'pearl' light bulbs sold for domestic use are not very satisfactory as they usually result in a 'hot-spot' of light in the centre of the field of view. An 'opal' bulb as used in photographic enlargers is much better. The stages in setting up the microscope are as follows.

(i)   Align the lamp and microscope, placing the lamp about 200–250 mm distant from the microscope. Set the lamp housing at a suitable angle and insert the largest field-limiting diaphragm or open the lamphouse iris fully, if fitted. Place the condenser at the top of its travel and open the condenser illuminating aperture diaphragm fully.

(ii)   Place a suitable contrasty specimen on the microscope stage and focus it with a 10× objective. There is usually sufficient light passing through the instrument for this to be done, even if the lamp and/or mirror are well out of alignment.

(iii)   Adjust the microscope mirror and lamp until the field of view is evenly illuminated.

(iv)   Move the mirror slightly so that the edge of the field-limiting diaphragm may be seen in the field of view. It will probably be unsharp, so adjust the position of the *condenser* up and down until its edge is in sharp focus. Because of the poor chromatic corrections of a typical Abbe condenser, this image may be fringed with colour; this is unavoidable. If field diaphragms are not available then the surface of the lamp bulb will have to suffice and be focused into the specimen plane. A pencil point placed in contact with the lamp surface will help to locate the point of focus.

(v)   Reduce the size of the field diaphragm until only the observed field is illuminated.

(vi)   Adjust the aperture iris below the condenser to give a suitable working aperture for the system (remove the eyepiece and inspect the back focal plane of the objective).

(vii)   Make minor adjustments to the mirror angle if necessary and use neutral density filters to reduce the intensity of the light if necessary; colour filters may be added if required to enhance the contrast of stained material.

7.9   Contrast enhancement

Although the importance of resolution of finer detail has been repeatedly stressed, an image must also have sufficient contrast to allow this detail to be seen easily. The introduction of contrast is, therefore, most important and many specialized optical techniques such as darkground, phase contrast and differential interference contrast are available for this purpose. Details of contrast enhancement have been covered in Bradbury and Evennett (1996).

If, however, we have a specimen which does not introduce much amplitude contrast then it is worth remembering that considerable contrast enhancement may be obtained very simply.

(i) The aperture diaphragm may be closed rather more than usual, so that the diffraction haloes so introduced emphasize the outlines of the structure. The finer details will be lost, but these would not be seen anyway if the contrast were low.

(ii) In conventional microscopy the light is arranged so that the focal point of the rays from the condenser is on the optical axis of the instrument; this is called axial illumination. By deliberately moving the condenser diaphragm (*not* the condenser itself) off-axis or by using an opaque substage stop with a hole towards its periphery we are able to obtain 'oblique' illumination. This technique often enhances structures so that they are seen clearly, but they appear in 'pseudo-relief' and their interpretation may thus be difficult (see *Figure 6.6b*).

If we have an object which is either naturally coloured or has been stained, it is possible to enhance or suppress contrast by using a suitable colour filter. This is especially valuable in monochrome photomicrography where the use of a filter of a complementary colour to that of the object will allow its contrast to be increased. For example, if a specimen is stained with dyes to give red and green areas then if it is photographed through a green filter the red areas in the print will be very dark whilst the green parts will be rendered much lighter (see *Figure 7.3*). Conversely, with a red filter the green areas will have increased contrast and appear dark whilst the red parts will be lighter and show detail.

Coloured filters, made either of glass or of gelatine, are available in a wide range of colours and spectral transmissions. Other filters of interest to the microscopist are colour-correction, neutral density and heat filters. Colour-correction filters are used to suppress the excess red and yellow in the light from tungsten lamps operated at low voltages. A light-blue filter may be used with advantage for this purpose even when using the microscope for visual work. For colour photomicrography colour-correction filters are used to alter the colour temperature of the light to suit the response of the film in use. For colour films balanced for artificial light these filters are pale blue but if daylight colour film is used then an appropriate filter (80B or similar) would be deep blue. The use of such filters is covered in detail in Bracegirdle and Bradbury (1995).

Neutral density filters (ND filters) do not alter the colour of the light but reduce its intensity. They are available in various densities from a very pale grey to almost opaque and they may be used in combination to attenuate the light as desired. If the lamp is not fitted with an intensity control, or if the colour temperature of the light must remain high, then ND filters are a useful way of reducing the brightness during visual observation. NEVER close the aperture diaphragm or lower the condenser from its true focus in order to reduce the brightness of the light!

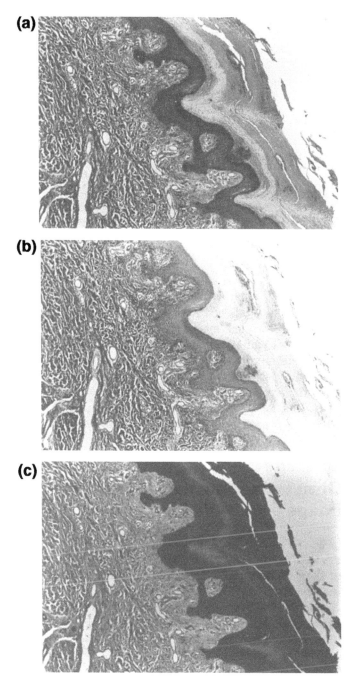

Figure 7.3. Contrast enhancement of coloured object by colour filters. The specimen is a stained section of skin where the keratinized layers (on the right) are coloured red. (a) Photograph in white light. All areas of the tissue show a strong contrast. (b) Photograph in red light. Both deep and superficial layers of the red-stained keratin on the right are now rendered much lighter in tone and detail is more easily visible. (c) Use of a green filter increases the contrast of the red-stained keratin layers so that they now appear almost black and without detail.

Table 7.1. Transmission of neutral density filters

Neutral density	Percentage transmission	Filter factor
0.1	80	$1^{1}/_{4}$
0.2	63	$1^{1}/_{2}$
0.3	50	2
0.4	40	$2^{1}/_{2}$
0.5	32	3
0.6	25	4
0.7	20	5
0.8	16	6
0.9	13	8
1.0	10	10
2.0	1	100
3.0	0.1	1000
4.0	0.01	10000

Neutral density filters are obtainable with 13 different values of absorption, as listed in *Table 7.1*.

If high-intensity lamps such as the tungsten–halogen types are used, a heat filter will normally be used in the light path of the lamp collector lens system. Heat filters are made of special glass designed to absorb much of the infrared radiation which causes heating of the specimen. They do, however, often introduce a greenish colour into the image which is not noticed when looking down the microscope but is very apparent with colour photomicrography. Do not remove any such filters since this may result in damage to either specimen or microscope components; counteract any colour cast by using suitable colour-compensating filters.

7.10 Dissatisfaction with the microscope image

If the microscope is correctly set up and the components are clean and undamaged then usually the image quality will be acceptable and the specimen may be studied with confidence. Since, unlike many other scientific instruments, the microscope will produce some sort of an image even if incorrectly adjusted, it is important to be able to recognize a poor image and understand the possible causes so that they may be rectified before time is wasted on the study of what may be no more than optical artifacts! Some of the possible causes of trouble along with their remedies are set out in *Table 7.2*.

Table 7.2. Poor image quality – a trouble-shooting guide

(1)	Is light present?	No	Continue
		Yes	Go to (6)
(2)	Is the microscope plugged in to a live socket and the mains switched on?	No	Rectify and go to (1)
		Yes	Continue
(3)	Has the bulb failed? Remove and test	No	Continue
		Yes	Replace and go to (1)
(4)	Is there an obstruction in the light path?	No	Ensure a low-power lens is in use
			Open all diaphragms fully
			Check that darkground stop or condenser is not in use
			Continue
		Yes	Remove obstruction and go to (1)
(5)	Is an external lamp and mirror in use?	No	Recheck all components and try again from (1)
		Yes	Misalignment most likely. Adjust and go to (1)
(6)	Is a blurred image present?	No	Continue
		Yes	Go to (10)
(7)	Is microscope wildly out of focus?	No	Continue
		Yes	Check that low-power (10×) objective is in use. Check contrasty specimen on stage. Refocus and replace with specimen for study. Go to (8)
(8)	Has specimen sufficient contrast for it to be plainly visible?	No	Try suitable enhancement
		Yes	Go to (9)
(9)	Is light too bright or not bright enough for specimen to be seen?	No	Continue
		Yes	Adjust with rheostat or by adding/removing ND filters
(10)	Is image present and in focus but field unevenly illuminated?	No	Continue
		Yes	Remedy and go to (19)
(11)	Is image present but drifts out of focus?	No	Continue
		Yes	Go to (23)
(12)	Does image appear hazy or have poor resolution? Do boundaries appear with haloes even when focus is adjusted?	No	Continue
		Yes	Go to (25)
(13)	Is image field fully illuminated or of inadequate size?	No	Continue
		Yes	Go to (36)
(14)	Is the image sharp over the entire field?	No	Go to (39)
		Yes	Continue

(Continued)

(Table 7.2 continued)

(15) Is the image sharp but of
low contrast?

No	Continue
Yes	Go to (43)

(16) Are there colour fringes
around contrasty objects?

No	Continue
Yes	Go to (50)

(17) Can the field diaphragm
be focused?

No	Go to (52)
Yes	Continue

(18) Are there shadows or
out-of-focus features in
the image?

No	Continue
Yes	Go to (55)

Uneven illumination

(19) Are areas of field obstructed,
especially at the edges?

Yes	Check for partial obstruction in the light path, e.g. nosepiece or filter holder incorrectly located
No	Continue

(20) Are there areas of greater
brightness (hot-spots) in
the field?

No	Go to (11)
Yes	Continue

(21) Is source-focused illumination
in use?

No	Go to (11)
Yes	(a) Check bulb for suitability
	(b) Use condenser with better corrections

(22) With Köhler illumination:
  (a) check alignment of
      lamp filament
  (b) check alignment and
      focus of lamp collector
      lens system
  (c) use condenser with
      better corrections

Then go to (11)

Focus drift

(23) If mountant is liquid,
is specimen moving?

Yes	(a) Allow mountant time to harden
	(b) Remove some of the fluid mountant
	(c) Allow time for specimen to settle
	(d) Do not use immersion lenses
No	Continue

(24) Probably a fault in the focus
mechanism. Have instrument
serviced

Go to (12)

(Continued)

(Table 7.2 continued)

Poor resolution

(25)	Is front surface of objective clean?	Yes	Continue
		No	Clean carefully (p. 118, Section 7.11)
(26)	If immersion lens in use, are there air bubbles in the oil?	Yes	Replace oil and make sure front element of objective and coverslip are in good contact with oil
		No	Continue
(27)	If high-aperture dry lens with correction collar is in use, is the collar correctly set for the coverslip thickness?	Yes	Continue
		No	Reset correction collar to correct value or use immersion objective
(28)	Is the mountant layer too thick?	Yes	Remount slide; use oil-immersion or dry objective with correction collar
		No	Continue
(29)	Is coverslip too thick to allow a high-aperture objective to focus on the specimen?	Yes	Remount with thinner coverslip or less mountant. Use objective with longer working distance
		No	Continue
(30)	With medium-power objectives: is slide upside-down on stage?	Yes	Invert slide
		No	Continue
(31)	Do edges of objects show pronounced haloes?	Yes	(a) Check that RI of mountant does not differ too much from that of specimen
			(b) Condenser aperture too small for the objective in use. Adjust
		No	Continue
(32)	Is field diaphragm open too far?	Yes	Glare causing image degradation. Readjust
		No	Continue
(33)	Is the objective NA too low to resolve detail required?	Yes	Change to higher NA and perhaps lower-power eyepiece
		No	Continue
(34)	Is the eyepiece magnification too high?	Yes	Change to higher-power objective and then use lower-magnification eyepiece
		No	Continue

(Continued)

(Table 7.2 continued)

(35)	Is the eyepiece magnification too low to allow visualization of detail?	Yes No	Replace with suitable eyepiece Go to (13)

Inadequate field of view

(36)	If coverage of object is adequate: is image of field diaphragm visible, even when fully open?	Yes No	Condenser of too short a focal length. Replace with different type or remove top lens Continue
(37)	Is coverage of object itself inadequate?	Yes No	(a) Use wide-field eyepiece or objective of lower magnification (b) Eyepiece exit pupil too close to eye lens for satisfactory viewing, especially if spectacles worn. Try high-eyepoint eyepiece Continue
(38)	With binocular head: is there a double image?	Yes No	(a) Interocular distance setting incorrect. Adjust (b) Eyepiece tube focus incorrect. Check by using each eye in turn and adjust if necessary Go to (14)

Image is unsharp in places

(39)	Is slide level on stage?	No Yes	Correct and continue Continue
(40)	With high-power objectives: does the sharpness of the image change from peripheral to central as focus is altered?	Yes No	Curvature of field effect. Use 'plan' or flat-field objective Continue
(41)	Does the orientation of image detail change as fine focus is altered?	Yes No	Marked astigmatism in objective. Try another Continue
(42)	Is the specimen too thick or poorly mounted?	Yes No	Use thinner specimen or lower-power objective. Improve mounting technique Go to (15)

(Continued)

(Table 7.2 continued)

Poor contrast

(43)	Has wrong mountant been used on low absorption specimen (i.e. is RI of mountant very close to that of specimen)?	Yes	Remount in medium of different RI	
		No	Continue	
(44)	Is field iris diaphragm opening too large?	Yes	Glare present, reducing contrast. Close to correct setting	
		No	Continue	
(45)	Is aperture diaphragm opening too large?	Yes	Glare present. Reset and continue	
		No	Continue	
(46)	Are all lens surfaces clean, especially objective front?	Yes	Continue	
		No	Clean carefully and continue	
(47)	If specimen is stained or coloured, is correct colour filter in use?	Yes	Continue	
		No	Change to complementary colour to increase contrast	
(48)	Is specimen of intrinsically low contrast?	Yes	Try staining or a contrast-enhancing technique such as darkground, phase contrast or DIC	
		No	Continue	
(49)	Is the light too bright?	Yes	Reduce intensity with rheostat or ND filters	
		No	Go to (16)	

Colour fringes

(50)	Are fringes only around image of field diaphragm?	Yes	Use achromatic condenser	
		No	Continue	
(51)	Are colour fringes present around edges of contrasty objects, especially near periphery of field?	Yes	(a)	Lateral chromatic difference of magnification present. Use correct eyepiece
			(b)	Poorly corrected achromat in use. Try a better-quality objective
			(c)	Use monochromatic light
		No	Go to (17)	

Poor image of field diaphragm

(52)	Is slide too thick or thin?	Yes	Replace if possible	
		No	Continue	
(53)	Is condenser correctly mounted and focused in substage holder?	Yes	Go to (18)	

(Continued)

(Table 7.2 continued)

	No	Insert correctly or focus and continue
(54) Is image of diaphragm not centred in field of view?	Yes	(a) Centre field iris with adjusting screws (b) Centre condenser with its screws if field iris not centrable
	No	Go to (18)

Out of focus features in image

(55) Do shadows appear and drift down through field of view?	Yes	Defects (*muscae volitantes*) in observer's eye
	No	Continue
(56) Do shadows or specks rotate if eyepiece is rotated?	Yes	Dirt on eye lens or field lens of eyepiece. Clean and continue
	No	Continue
(57) Do shadows or specks alter in sharpness as condenser is moved up and down?	Yes	(a) Dirt on collector lens or filter. Clean and continue (b) With source focus: marks or dirt on front surface of bulb (c) Image of ground glass in optical system. Remove if possible
	No	Continue
(58) Do shadows or specks move with specimen?	Yes	(a) Dirt on or in slide. Remount if possible (b) Possible refraction from e.g. air bubbles in mountant. Remount if possible

7.11   Care of the instrument

Full servicing of the microscope is best left to a competent service engineer. There are, however, some tasks which the user may perform and which if carefully done will help keep the instrument in top condition. As with any precision instrument, prevention is better than cure, and so it is important to protect the microscope and its components from mechanical damage. Do not unscrew the objectives from a nosepiece unless it is essential since, when changing objectives, it is easy to drop one on to the bench should the screw thread not immediately engage. Accidents such as this may be prevented by holding the objective in place in the nosepiece aperture with the fingers of one hand around it whilst screwing it home with the other hand. Only let go with the holding fingers when you are absolutely sure that the objective thread is firmly engaged in the nosepiece. Again, if you remove eyepieces make sure that they are placed securely so that they cannot roll off the bench on to the floor.

Optical glass and its coatings may be very soft and therefore easily scratched by minute particles of grit which settle on them from atmospheric dust. For this reason the instrument should not remain exposed; plastic dust covers are cheap and should always be used when the microscope is not in use for any length of time. Never leave the eyepiece tubes empty or spaces in the nosepiece without an objective. Blanking caps are available for both of these if the lenses have to be removed for any extended period. Again, grit may easily become embedded in cleaning cloths, so whenever optical surfaces are to be cleaned *only lens tissues* should be used, taken fresh from the packet each time. The other major enemy of optical surfaces is grease. This may come from an accidental touch with a finger. The front lens of a 'dry' objective may easily become contaminated with immersion oil or mountant. This will cause very poor contrast and image resolution and often the appearance of the image in such cases simulates the effect seen when gross spherical aberration is present (see *Figure 1.14*).

Dust on the surface of lenses should in the first instance be removed by gentle air blasts from a puffer or by a gentle squirt from one of the containers of compressed gas now available as laboratory dusters. More persistent dirt and grease on the lenses may be removed by gentle wiping with a fresh lens tissue moistened with a proprietary lens-cleaning fluid. The use of excess fluid should be avoided and the operation ended by careful wiping with a fresh dry tissue.

The use of ethanol is not recommended as it may loosen the cement holding some of the front elements of the lenses: iso-propanol is usually suitable. Under no circumstances should you try to dismantle objectives; they can only be reassembled correctly by the makers who use special jigs for this purpose. Some of the older and simpler eyepieces and condensers may be dismantled for cleaning provided that care is taken to note the sequence and orientation of the various lenses as they are removed. Most are held by retaining rings with very fine-pitch screw threads, so be careful on reassembly to ensure that they are not cross-threaded. Never use excessive force on any component either to dismantle or to reassemble it. If in doubt, *always* seek competent professional help.

The mechanical parts of modern microscopes require very little other than an occasional wipe to remove dust. Most of the rackwork is now completely enclosed, and the stage and focusing movements should not require any attention from users. The stage may be subject to soiling from liquid mountants: if any is spilled, it should immediately be wiped away with soft tissue before it dries and possibly causes corrosion.

The electrical components of the microscope should need no routine maintenance from the user. It is assumed that in a professional laboratory they will have passed the inspection of the safety officer responsible for all electrical equipment. Bulbs, of course, must be replaced as and when they fail. If tungsten–halogen bulbs are in use remember that these should never be handled with bare hands. Grease contamination from the fingers on the quartz envelope of a new bulb could materially shorten its working life, so handle new bulbs with tissue when fitting them into their holders. Make sure that the lamp housing is properly fastened before using the lamp. This is especially important with mercury discharge tubes as they contain vapour under high pressure.

Do-it-yourself attempts at servicing may sometimes be very ill-advised and could seriously affect the subsequent performance of the microscope. Expensive damage may be caused through ignorance so the golden rule is: IF IN DOUBT – LEAVE WELL ALONE. It is always worthwhile calling in the professional service engineer rather than risking damage to a valuable instrument.

Reference

Bracegirdle, B. and Bradbury, S. (1995) *Modern PhotoMICROgraphy.* Royal Microscopical Society Handbook No. 33. BIOS Scientific Publishers, Oxford.

Bibliography

Bradbury, S. (1992) *Basic Measurement Techniques for Light Microscopy.* Oxford University Press and Royal Microscopical Society, Oxford.

Bradbury, S. and Evennett, P. (1996) *Contrast Techniques in Light Microscopy.* BIOS Scientific Publishers, Oxford.

Bradbury, S., Evennett, P.J., Haselmann, H. and Piller, H. (1989) *RMS Dictionary of Light Microscopy.* Oxford University Press and Royal Microscopical Society, Oxford.

Delly, J.G. (1988) *Photography Through the Microscope.* Eastman Kodak Company, Rochester, NY.

Gifkins, R.C. (1970) *Optical Microscopy of Metals.* Pitman, Melbourne.

Hartley, W.G. (1993) *The Light Microscope – Its Use and Development.* Senecio Publishing Company, Oxford.

Oldfield, R. (1994) *Light Microscopy, An Illustrated Guide.* Wolfe, Mosby-Yearbook, Europe, London.

Ploem, J.S. and Tanke, H.J. (1987) *Introduction to Fluorescence Microscopy.* Oxford University Press and Royal Microscopical Society, Oxford.

Pluta, M. (1988) *Advanced Light Microscopy. Vol. 1 – Principles and Basic Properties.* Polish Scientific Publishers/Elsevier, Amsterdam.

Pluta, M. (1989) *Advanced Light Microscopy. Vol. 2 – Specialized Methods.* Polish Scientific Publishers/Elsevier, Amsterdam.

Pluta, M. (1993) *Advanced Light Microscopy. Vol. 3 – Measuring Techniques.* Polish Scientific Publishers/Elsevier, Amsterdam.

Richardson, J.H. (1991) *Handbook for the Light Microscope: A User's Guide.* Noyes Publications, New Jersey, USA.

Robinson, P.C. and Bradbury, S. (1992) *Qualitative Polarized-Light Microscopy.* Oxford University Press and Royal Microscopical Society, Oxford.

Slayter, E.M. and Slayter, H.S. (1993) *Light and Electron Microscopy.* Cambridge University Press, Cambridge.

Smith, R.F. (1994) *Microscopy and Photomicrography, A Working Manual.* CRC Press, Boca Raton, FL.

Spencer, M. (1982) *Fundamentals of Light Microscopy.* Cambridge University Press, Cambridge.

Thomson, D.J. and Bradbury, S. (1987) *An Introduction to Photomicrography*. Oxford University Press, Oxford.

The following books may also be useful. Some have long been out of print, but they may be found in libraries and on the second-hand market.

Allen, R.M. (1940) *The Microscope*. Van Nostrand, Princeton, NJ.

Barron, A.L.E. (1965) *Using the Microscope*, 3rd Edn. Chapman & Hall, London.

Beck, C. (1923) *The Microscope*, 2nd Edn. R. & J. Beck Ltd, London.

Carpenter, W.B. and Dallinger, W.H. (1891, 1901) *The Microscope and its Revelations*, 7th and 8th Edns. Churchill, London.

Gage, S.H. (1941) *The Microscope*, 17th Edn. Comstock Publishing Co. (Constable), London.

Hartley, W.G. (1979) *Hartley's Microscopy*. Senecio Press, Oxford.

Hogg, J. (1898) *The Microscope, its History, Construction and Application*, 15th Edn. George Routledge and Sons Ltd, London.

McCrone, W.C., McCrone, L. and Delly, J.G. (1978) *Polarized Light Microscopy*. Ann Arbor Science, MI.

Needham, G.H. (1958) *The Practical Use of The Microscope*. Thomas, Springfield, IL.

Payne, B.O. (1957) *Microscope Design and Construction*. Cooke, Troughton & Simms, York.

Spitta, J. (1909, 1920) *Microscopy – The Construction, Theory and Use of the Microscope*, 2nd and 3rd Edns. John Murray, London.

Index

Milton Keynes UK
Ingram Content Group UK Ltd.
UKHW031152141024
449569UK00024B/862